Springer Proceedings in Materials

Volume 21

AF904608

Series Editors

Arindam Ghosh, Department of Physics, Indian Institute of Science, Bangalore, India

Daniel Chua, Department of Materials Science and Engineering, National University of Singapore, Singapore, Singapore

Flavio Leandro de Souza, Universidade Federal do ABC, Sao Paulo, São Paulo, Brazil

Oral Cenk Aktas, Institute of Material Science, Christian-Albrechts-Universität zu Kiel, Kiel, Schleswig-Holstein, Germany

Yafang Han, Beijing Institute of Aeronautical Materials, Beijing, Beijing, China

Jianghong Gong, School of Materials Science and Engineering, Tsinghua University, Beijing, Beijing, China

Mohammad Jawaid, Laboratory of Biocomposite Technology, INTROP, Universiti Putra Malaysia, Serdang, Selangor, Malaysia

Springer Proceedings in Materials publishes the latest research in Materials Science and Engineering presented at high standard academic conferences and scientific meetings. It provides a platform for researchers, professionals and students to present their scientific findings and stay up-to-date with the development in Materials Science and Engineering. The scope is multidisciplinary and ranges from fundamental to applied research, including, but not limited to:

- Structural Materials
- Metallic Materials
- Magnetic, Optical and Electronic Materials
- Ceramics, Glass, Composites, Natural Materials
- Biomaterials
- Nanotechnology
- Characterization and Evaluation of Materials
- Energy Materials
- Materials Processing

To submit a proposal or request further information, please contact one of our Springer Publishing Editors according to your affiliation:

European countries: **Mayra Castro** (mayra.castro@springer.com)
India, South Asia and Middle East: **Priya Vyas** (priya.vyas@springer.com)
South Korea: **Smith Chae** (smith.chae@springer.com)
Southeast Asia, Australia and New Zealand: **Ramesh Nath Premnath** (ramesh.premnath@springer.com)
The Americas: **Michael Luby** (michael.luby@springer.com)
China and all the other countries or regions: **Mengchu Huang** (mengchu.huang@springer.com)
This book series is indexed in **SCOPUS** database.

Shen-Ming Chen
Editor

Proceedings of 10th International Conference on Chemical Science and Engineering

ICCSE 2021

Springer

Editor
Shen-Ming Chen
National Taipei University of Technology
Taipei, Taiwan

ISSN 2662-3161 ISSN 2662-317X (electronic)
Springer Proceedings in Materials
ISBN 978-981-19-4289-1 ISBN 978-981-19-4290-7 (eBook)
https://doi.org/10.1007/978-981-19-4290-7

© Springer Nature Singapore Pte Ltd. 2023
This work is subject to copyright. All rights are solely and exclusively licensed by the Publisher, whether the whole or part of the material is concerned, specifically the rights of translation, reprinting, reuse of illustrations, recitation, broadcasting, reproduction on microfilms or in any other physical way, and transmission or information storage and retrieval, electronic adaptation, computer software, or by similar or dissimilar methodology now known or hereafter developed.
The use of general descriptive names, registered names, trademarks, service marks, etc. in this publication does not imply, even in the absence of a specific statement, that such names are exempt from the relevant protective laws and regulations and therefore free for general use.
The publisher, the authors, and the editors are safe to assume that the advice and information in this book are believed to be true and accurate at the date of publication. Neither the publisher nor the authors or the editors give a warranty, expressed or implied, with respect to the material contained herein or for any errors or omissions that may have been made. The publisher remains neutral with regard to jurisdictional claims in published maps and institutional affiliations.

This Springer imprint is published by the registered company Springer Nature Singapore Pte Ltd.
The registered company address is: 152 Beach Road, #21-01/04 Gateway East, Singapore 189721, Singapore

Organization

Conference Chair

Prof. Zongjin Li, University of Macau (UM), China

Conference Co-chair

Prof. Shen-Ming Chen, National Taipei University of Technology, Taiwan

Program Chairs

Prof. Kwang Leong Choy, University College London, UK
Prof. Ashok Srivastava, Louisiana State University, USA
Assoc. Prof. Yulong Sun, Northwestern Polytechnical University, China

Publicity Chair

Prof. Parames Chutima, Chulalongkorn University, Thailand

Technical Program Chairs

Behzad Nematollahi, Swinburne University of Technology, Australia
Changxue Xu, Texas Tech University, USA

Eduardo M. G. Rodrigues, Management and Production Technologies of Northern Aveiro, Portugal
Faruk Elaldi, University of Baskent, Turkey
Guohua Xie, Wuhan University, China
Hongbo Zhang, East China University of Science and Technology, China
Khaled Abou-El-Hossein, Nelson Mandela University, South Africa
Marcelo H. Prado da Silva, Military Institute of Engineering-IME, Brazil
Nguyen Thanh Tu, Nguyen Tat Thanh University, Vietnam
Pavel Ripka, Czech Technical University, Czech Republic
Qifeng Zhang, North Dakota State University, USA
Rattanaphol Mongkholrattanasit, Rajamangala University of Technology Phra Nakhon, Thailand
Ren Jianxin, Lanzhou Jiaotong University, China
Ricardo Branco, University of Coimbra, Portugal
Sheila Shahidi, Islamic Azad University, Iran
Siripron Sripiboon, Rangsit University, Thailand
Songling Huang, Tsinghua University, China
Subbiah Alwarappan, CSIR-Central Electrochemical Research Institute Karaikudi, India
Supawan Tirawanichakul, Prince of Songkla University, Thailand
Suriati binti Sufian, Universiti Teknologi PETRONAS Malaysia, Malaysia
Tingkai Zhao, Northwestern Polytechnical University, China
V. A. Brodskiy, D. Mendeleev University of Chemical Technology of Russia, Russia
Vu Ngoc Pi, Nguyen Tat Thanh University, Vietnam
Xiaoliang Zhu, Hitachi America Ltd. R&D, China
Yang Xi, II-VI Corp., USA
Yu-Chung Chang, National Changhua University of Education, Taiwan

Preface

2021 10th International Conference on Chemical Science and Engineering (ICCSE) was held online during November 19–21, 2021, the conferences were great success.

Due to the outbreak of COVID-19, this year's conference, which was supposed to be held in University of Macau, China, was organized as a virtual conference, using of Zoom platform. It was a fully synchronized online conference that had live Q&A, participants jointly listened to presentations, and attended other live events associated with the conference. The dates of the virtual conference remained unchanged to the initially proposed dates, which were on November 19–21, 2021.

The conference program was organized in 7 sessions, in which 5 plenary and 49 specialized reports were presented. The scientists from the China, Korea, Malaysia, Thailand and others, took part in the scientific forum. Each speech lasted for 40 m, including 5 m for Q&A, and for other sessions, there was 15 m for each presentation, including 2 m for Q&A. The interaction session was made the real experience to the participants and they had good exposure with the foreign university students and delegates.

The major goal of ICCSE is to bring academic scientists, engineers, and industry researchers together to exchange their experiences and research results, and discuss the practical challenges they encountered and the solutions they adopted. All the papers in this proceeding were subject to peer-review by conference committee members and international reviewers. The papers were selected for the proceedings based on their quality and their relevance to the conference.

We would like to thank all the author submitted papers to the conferences and thank you all the reviewers for their time and effort in reviewing articles. Finally, it is a great honor to thank you all of behind staff who made their constant effort to make sure the conference process can go smoothly and the proceeding team who have dedicated their constant support and countless time to bring these scratches into this volume.

Taipei, Taiwan | Prof. Shen-Ming Chen

Contents

Polymer Chemistry and Engineering

Model-Based Predictive Controller for a Polylactic Acid Ring-Opening Polymerization Process

Thitiworada Rattanagorn, Wachira Daosud, and Paisan Kittisupakorn

1 Introduction

Reducing the amount of plastic waste is one of the most concerned environmental topics nowadays. Unfortunately, campaigns against plastic usage are not successful as it should be as global plastic market size is continually increasing. As a result, biodegradable plastic, which is one of the most prominent alternative solutions, has gotten a lot of attention in recent years. And among those plastics, PLA, due to its flexibility, relatively easy to produce, low toxicity and environmentally friendly, is one the most widely used biodegradable thermoplastics in the market, especially in packaging and medical device industries [1–3].

In general, there are three methods of synthesizing polylactic acid. The first and most straightforward method is direct polycondensation polymerization. However, the downside of this method is that it is generally unable to produce high molecular weight polymers due to the formation of water in the reaction. The second method is azeotropic condensation polymerization which adds distillation and catalytic process to remove water. However, these add complexity to the process and can adversely affect PLA biocompatibility. For those reasons, the trendiest method used nowadays and the one used in this research is ring open polymerization (ROP) which is a more complex method that is able to produce both high molecular weight polymer and very pure PLA.

T. Rattanagorn · P. Kittisupakorn (✉)
Control and Systems Engineering Research Laboratory, Department of Chemical Engineering, Faculty of Engineering, Chulalongkorn University, Bangkok 10330, Thailand
e-mail: paisan.k@chula.ac.th

W. Daosud
Department of Chemical Engineering, Faculty of Engineering, Burapha University, Chonburi 20130, Thailand

© The Author(s), under exclusive license to Springer Nature Singapore Pte Ltd. 2023
S.-M. Chen (ed.), *Proceedings of 10th International Conference on Chemical Science and Engineering*, Springer Proceedings in Materials 21,
https://doi.org/10.1007/978-981-19-4290-7_1

As for control, model predictive control (MPC) is used as the main technique in this research. The principle of this model-based control is to calculate a set of future control action by minimizing a specified objective function. As the controller directly use its process model, it is able to provide high control performance as long as the model reasonably represents the process which resulted in it is being one of the most widespread advanced process controls in the industry [4].

As a results, this paper presents and discusses the performance tests between MPC controls and optimized PID controls which are the type of control schemes currently used in the industry. The control performance characteristics used for comparison in this paper are the rise time, percentage overshoot and integral absolute error (IAE). The batch reactor model of the PLA process was simulated using MATLAB.

2 Process Description and Simulation

2.1 *Kinetic Modeling*

In this research, the elementary kinetic scheme of PLA formation [5] is enhanced with transesterification and non-radical random chain scissions for all relevant reaction stages to provide more accurate molecular weight and conversion factor [6, 7]. This process uses $Sn(Oct)_2$ as the catalyst and 1-dodecanol as co-catalyst.

To represent the dynamics of the chain length distributions and the molecular weight distribution of PLA. The values are estimated using a statistical method, modifying the three chain populations to nine moment balances (active, dormant and dead respectively indicated with λ, μ and γ). The conversion factor is calculated from Eq. (1). The number average molecular weight (*Mn*) and weight average molecular weight (*Mw*) are calculated from Eq. (2) and Eq. (3), respectively.

$$X = 1 - \frac{[M]}{[M_0]} \tag{1}$$

$$Mn = \frac{\lambda_1 + \mu_1 + \gamma_1}{\lambda_0 + \mu_0 + \gamma_0} M_{\text{mon}} \tag{2}$$

$$Mw = \frac{\lambda_2 + \mu_2 + \gamma_2}{\lambda_1 + \mu_1 + \gamma_1} M_{\text{mon}} \tag{3}$$

where $[M]$ and $[M_0]$ are the concentration of monomer, $\lambda_0, \lambda_1, \lambda_2$ are zero, first and second moments of active chains, respectively. μ_0, μ_1, μ_2 are zero, first and second moments of dormant chains, respectively. $\gamma_0, \gamma_1, \gamma_2$ are zero, first and second moments of dead chains, respectively. M_{mon} is the molecular weight of lactide monomer.

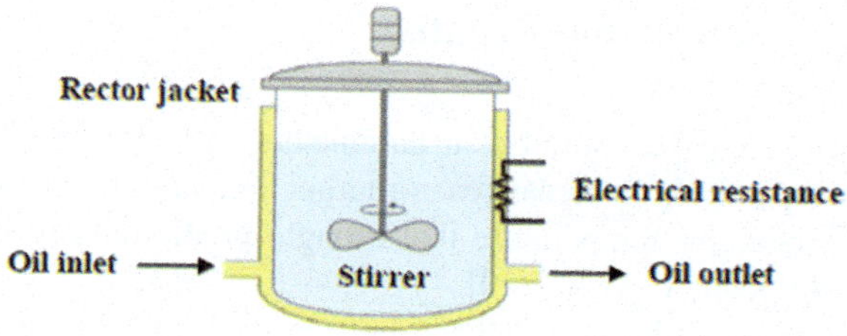

Fig. 1 Batch reactor configuration

2.2 *Batch Reactor Model*

The reactor configuration for this study simulation is based on a stirred batch reactor from a pilot plant operating at BIOFABRIS, University of Campinas, for poly(methyl methacrylate) (PMMA) synthesis [8]. The polymerization is an exothermic process and is processed at atmospheric pressure. Reactor temperature is chosen as the control variable as it is a primary variable directly affecting the yield of the process and is easily manipulated [8]. To control the temperature, diathermic oil is fed pumped inside the jacket at room temperature and electrical resistance is installed in the jacket to increase the temperature of the oil if necessary, as shown in Fig. 1.

A heat transfer term of reactor and reactor jacket energy balance was added to develop the complete reactor model.

$$m_{\mathrm{TOT}} C_p \frac{\mathrm{d}T}{\mathrm{d}t} = [(-\Delta H_{r,a}) r_a + (-\Delta H_{r,p}) r_p] V_r + \mathrm{UA}(T_j - T) \tag{4}$$

$$\rho_{\mathrm{oil}} V_j C_{p,\mathrm{oil}} \frac{\mathrm{d}T_j}{\mathrm{d}t} = Q_{\mathrm{oil}} \rho_{\mathrm{oil}} C_{p,\mathrm{oil}} (T_{j,0} - T_j) + \mathrm{UA}(T - T_j) + pSP_e \tag{5}$$

The jacket temperature is adjusted by the normalized power value (from 0 to 1) of electrical power resistance so that when the signal of 0 is applied no electrical power is supplied to the jacket and when the signal of 1 is applied the maximum value is supplied, whereas pS is the normalized power factor, P_e is the maximum electrical power resistance available (20 kW) and the oil flux fed to the jacket is 130 L/h.

As the reaction volume is not constant during polymerization, the dilution term to the material and moment balances must be added as follow:

$$\frac{\mathrm{d}C_i}{\mathrm{d}t} = r_i - \frac{C_i}{V_r} \frac{\mathrm{d}V_r}{\mathrm{d}t} \tag{6}$$

where C_i is the concentration of the generic species i and r_i is the global rate of formation.

The ODE model was developed in MATLAB and solved using ode45 solvers which provide dynamic prediction results of *X*, *Mn*, *Mw*, reactor temperature, jacket temperature and liquid level.

3 Simulation Results

In this paper, simulation parameters M/C and ROH/C were set as 6000 and 15, respectively, whereas reactor target temperature was set at 200 °C which is the best productive temperature that brought prediction results to the process goal (Mw = 100 kDa, $X = 95\%$) [9].

As for the performance indices, rise time (*t*r), percentage overshoot (MP) and integral of absolute error (IAE) are utilized for the performance comparison between the MPC controller and the optimized PID controller.

Since the K_D derivative is not necessary for the response of this process, PI control was developed, and the control parameters were selected to minimize oscillations of both controlled and manipulated variables. As a result, optimal values obtained for K_P and K_I are 0.18 and 0.20, respectively. On the other hand, for MPC controls, values of the control interval, prediction horizon, control horizon, input weight and output weight were chosen as 0.6, 5, 2, 10 and 1, respectively.

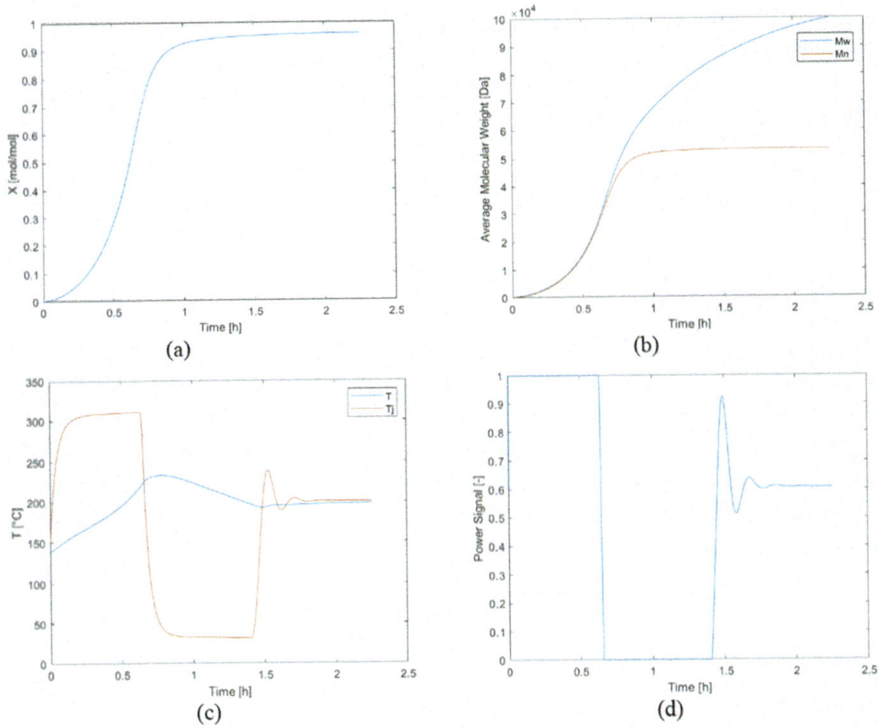

Fig. 2 Temperature control effect of the PI controller, **a** X with *time*, **b** Mn and Mw with *time*, **c** T and T_j with *time* and **d** pS with *time*

The system response of the PI controller and the MPC is shown in Figs. 2 and 3, and the results of the performance measurements of both controls are shown in Table 1.

As shown in Table 1, the MPC is superior to the PI controller in all indicators.

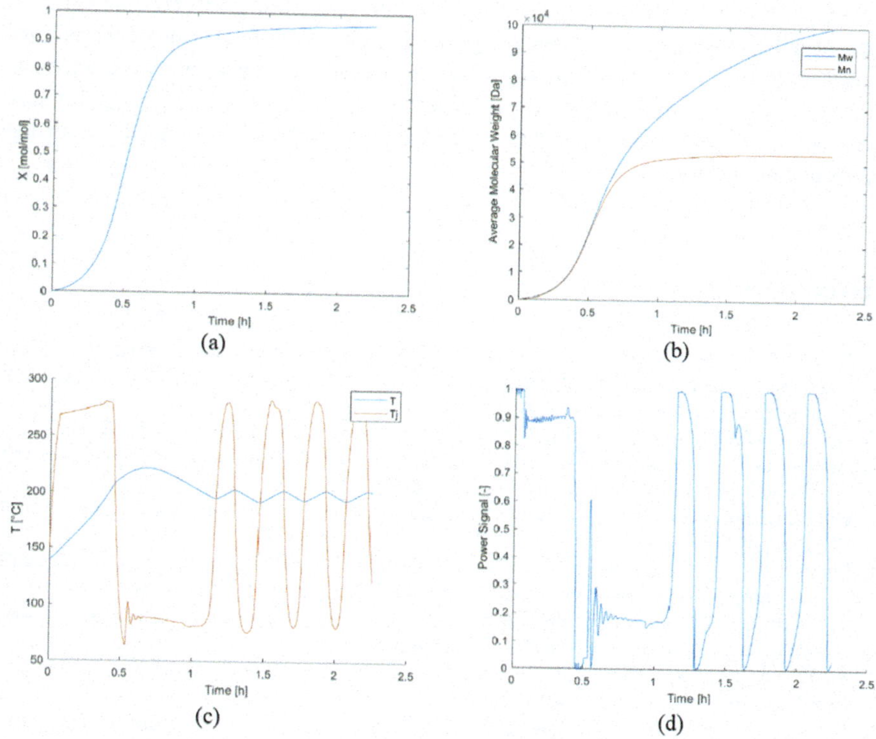

Fig. 3 Temperature control effect of the MPC controller, **a** *X* with *time*, **b** *Mn* and *Mw* with *time*, **c** T and T_j with *time* and **d** *pS* with *time*

Table 1 Comparison of control performance between PI and MPC controller

Performance indicators	Controller	
	PI	MPC
t_r (h)	0.438	0.429
MP (%)	26.508	26.471
IAE	39.3637	34.549

4 Conclusion

The study presents effective methods to control process conditions to produce high polymer quality in biomedical applications ($Mw = 100{,}000$ Da and residual monomer content 5%) within complete polymerization time of 2.2 h without causing any degradation phenomena. Performance comparison between MPC and PID controllers is measured based on rise time, percentage overshoot and IAE. Although the control effect of the MPC is quite oscillating which is believed to be caused by the selection of prediction horizon, both PI and MPC control can satisfy the main control objectives. And, the MPC controller performed better than the PID controller in all the performance indicators.

References

1. Dubois P, Jacobs C, Jérôme R, Teyssie P (1991) Macromolecular engineering of polylactones and polylactides. 4. Mechanism and kinetics of lactide homopolymerization by aluminum isopropoxide. Macromolecules 24(9), 2266–2270
2. Garlotta D (2001) A literature review of poly (lactic acid). J Polym Environ 9(2):63–84
3. Abd Razak SI, Ahmad Sharif NF, Abdul Rahman WAW (2012) Biodegradable polymers and their bone applications: a review. Int J Basic Appl Sci 12:31–49
4. Qin SJ, Badgwell TA (2003) A survey of industrial model predictive control technology. Control Eng Pract 11(7):733–764
5. Mehta R, Kumar V, Upadhyay SN (2007) Mathematical modeling of the poly (lactic acid) ring–opening polymerization kinetics. Polym-Plast Technol Eng 46(3):257–264
6. Yu Y, Storti G, Morbidelli M (2009) Ring-opening polymerization of L, L-lactide: kinetic and modeling study. Macromolecules 42(21):8187–8197
7. Yu Y, Storti G, Morbidelli M (2011) Kinetics of ring-opening polymerization of l, l-lactide. Ind Eng Chem Res 50(13):7927–7940
8. Linan LZ, Bonon A, Lima NM, Maciel R (2013) Quality control of Poly (methyl methacrylate) to medical purpose by multiple headspace extraction. Chem Eng 32
9. Refinettia D, Sonzognia A, Manenti F, Nascimento NM, Limab LZL, Filhob RM (2014) Modeling and simulation of poly (l-lactide) polymerization in batch reactor. Chem Eng 37

Micro-Fibrillated Cellulose Prepared from Sorghum Bicolor (L.) Moench by TEMPO-Mediated Oxidation Treatment

Achmad Nandang Roziafanto, Muhammad Furqon, Nofrijon Sofyan, and Mochamad Chalid

1 Introduction

One of the most prevalent natural polymers is cellulose which is renewable, biodegradable, biocompatible and low-cost material [1, 2]. Many researchers are concentrating their efforts on developing methods for extracting MFC from natural fibers like flax, hemp, jute, sisal, sugar palm/ijuk "Arenga pinnata," empty fruit bunches of oil palm and kenaf [3–5]. MFC from natural fiber is a promising renewable material to produce an environmentally friendly composite material [3, 6]. Their stiffness properties and their great crystallinity make them a potential for reinforcement in composite [1, 2]. Sorghum (Sorghum bicolor (L.) Moench) is the fifth most cultivated crop in the world and its stalks contain 40–49% cellulose, 17–25% lignin and 27–32% hemicellulose [7]. Because of its high cellulose contains, sorghum will be a potential resource to produce MFC as a polymer composite reinforcement to enhance its performance and crystallinity [8, 9].

The MFC extraction from natural fibers in principle must remove hemicellulose and lignin that are connected with cellulose in chemical bonding through a biological, physical, mechanical and chemical treatment to break the bonding. Among all treatments for MFC extraction, chemical treatment is well known and more simple and cheaper. From our previous study, we obtained MFC from sorghum with high crystallinity index of around 75,73% using alkalinization with sodium hydroxide followed by bleaching with sodium chlorite with a triplicate of graduation treatment, but they are limitless in time-consuming processes and less effective [8]. Alternative

A. N. Roziafanto · M. Furqon · N. Sofyan · M. Chalid (✉)
Department of Metallurgical and Material, Faculty of Engineering, Universitas Indonesia, Depok, Indonesia
e-mail: chalid@metal.ui.ac.id

A. N. Roziafanto
Politeknik AKA Bogor, Ministry of Industry Indonesia, Bogor, Indonesia

© The Author(s), under exclusive license to Springer Nature Singapore Pte Ltd. 2023

S.-M. Chen (ed.), *Proceedings of 10th International Conference on Chemical Science and Engineering*, Springer Proceedings in Materials 21,
https://doi.org/10.1007/978-981-19-4290-7_2

methods with catalytic oxidation of natural fiber using NaClO and 4-H-TEMPO and $NaClO_2$ as a primary oxidant were more effective to convert the fiber onto individual fibrils with uniform size [10]. These treatments can be more efficient and beneficial by utilizing low-energy approaches. The purpose of this study is to evaluate the effect of a graduation treatment of alkalinization followed by TEMPO-mediated oxidation to MFC sorghum properties. The graduation treatment was aimed to defibrillate the natural fiber into microfibrils with a high crystallinity ratio. The obtained MFC was characterized using FTIR spectroscopy, FE-SEM and XRD to evaluate their chemical composition, morphology and crystallinity index.

2 Materials and Methods

2.1 *Material*

Sorghum fiber stalks were sourced at the traditional market in Bogor, Indonesia. Sorghum fibers were dried at 30–35 °C for three days and crushed into 40 mesh. The 2,2,6,6-tetramethylpiperidine-1-oxyl (4-H-TEMPO), NaOH, $NaOCl_2$, NaOCl and H_2SO_4 were purchased from Merck and used without further purification.

2.2 *MFC Preparation*

In this study, seven grams of sorghum fibers (SF) 40 mesh were subjected to alkalinization treatment (SF A) and TEMPO-mediated oxidation treatment (SF A-T). The alkalinization treatment (SF A) was produced by immersing sorghum fibers in 20% (w/w) NaOH solution with continuous agitation for 2 h at 80 °C. For 1 h, the mixture was maintained at 80 °C. After alkalinization, a TEMPO-mediated oxidation treatment (SF A-T) was performed by soaking in a solution composed of 4-H-TEMPO (0.1 mmol), $NaOCl_2$ (10 mmol) and 0.1 M acetate buffer (100 mL, pH 4.8). In a single step, a 2 M NaOCl solution (0.5 mL, 1.0 mmol) was added to the flask. The flask was immediately sealed, and the suspension was stirred at 50 °C for 2 h (SF A-T2h) or 6 h (SF A-T6h). Following each sequence treatment, samples were washed three times with deionized water and submerged for 1 h in deionized water until the pH reached about 7. For characterization, the washed fibers were dried at room temperature and stored in a dry container.

2.3 *MFC Characterization*

Characterization of chemical composition, morphology, and crystallinity using Infrared Spectrometer FTIR Spectrum TwoTM Perkin Elmer (ASTM E 1252), FE-SEM Inspect F50, and X-Ray Diffractometer Shimadzu XRD-7000 instruments. The FE-SEM analysis was used to determine the microstructure transformation of sorghum fibers after treatment. The crystallinity of untreated and treated sorghum was determined using XRD and FTIR. The crystallinity was determined by Eq. 1 adopted from the method of Segal et al. [11].

$$CI = \frac{I_{002} - I_{am}}{I_{002}} \times 100\% \quad (1)$$

where CI denotes the crystallinity index, I (002) is the highest intensity at 2 = 22° and I(amp) denotes the amorphous intensity scattered at 2 = 18°.

3 Results and Discussion

3.1 *Chemical Composition in Fiber*

Chemical alkalization followed by TEMPO-mediated oxidation of MFC derived from sorghum fibers might alter the chemical composition in sorghum fibers. We use FTIR to determine the chemical composition of sorghum fibers untreated and treated by alkalinization and TEMPO-mediated oxidation, as shown in Fig. 1.

The peaks on the spectrum in Fig. 1 show the absorbance peak for the different chemical bonds found in untreated and chemically treated sorghum fibers. As shown in Fig. 1, the absorption around 1000–1300 cm^{-1} in all-fiber samples corresponds to C-O stretch in hemicellulose and lignin [3, 6, 9, 12] and absorption in wavenumber 1500–1600 cm^{-1} corresponds to C–C bond in aromatic lignin [9]. Another absorption that appeared between 3600–3000 cm^{-1} is associated with the –OH stretch in cellulose and/or hemicellulose [8, 9, 13]. The other absorption band between 2930 and 2910 cm^{-1} is connected to the CH stretch vibration of the carbonyl aldehyde groups in lignin [3, 6, 9, 12]. The absorption at 1430 cm^{-1} corresponds to the vibration of symmetric CH2, also known as the cellulose crystallinity zone. The absorption at 904–893 cm^{-1} is due to the C–O–C bond in β (1–4) glycosidic ether, also known as cellulose's amorphous area [8].

This study compared untreated sorghum fiber to fibers treated with alkalinization and alkalinization followed by TEMPO-mediated oxidation to achieve MFC by removing contaminants such as wax and oil that obscure the fiber's outer surface, as well as hemicellulose and lignin. The FTIR result for sorghum fibers treated with alkalinization/TEMPO-mediated oxidation resulted in a decline in intensity at the

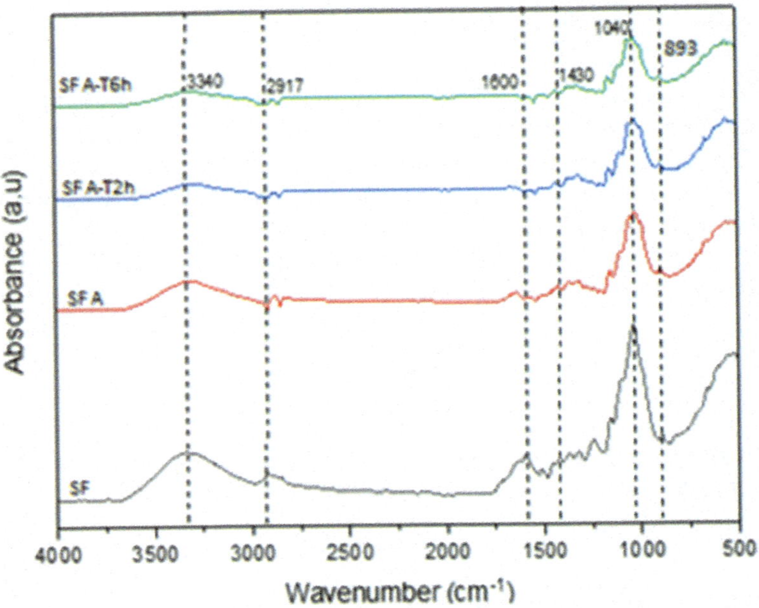

Fig. 1 FTIR spectra of sorghum untreated (SF) and treated by Alkalinization (SF A) and TEMPO-mediated oxidation (SF A-T2h; SF A-T6h)

wavenumber around 1000–1300 cm^{-1} and 1600 cm^{-1}, but there is no significant difference in its intensity for 2 h and 6 h treatment. This indicates that some lignin and hemicellulose are removed from sorghum fibers during alkalinization and that further lignin and hemicellulose are removed during TEMPO-mediated oxidation treatment. These findings suggest that alkalinization/TEMPO-mediated oxidation is the most effective method for removing binding compounds such as lignin, hemicellulose and pectin from fibers. When that bonding substance is removed, the fiber bundle defibrillates [3, 8]. Additionally, alkalinization/TEMPO-mediated oxidation was able to raise the crystallinity ratio in sorghum fiber, as demonstrated by a rise in absorbance intensity between 904 and 893 cm^{-1} and a decrease in absorbance intensity between 1430 and 1440 cm^{-1}. This implies that the treatment had eliminated some of the amorphous regions. The alkalinization/TEMPO-mediated oxidation has great performance in eliminating undecided compounds in the sorghum and eliminating the amorphous region in cellulose compare to alkalinization.

Absorption peaks in regions 1700–850 cm^{-1} can be useful for studying the polymorph of cellulose. The absorption peak assignments for each sample are shown in Table 1 which is compared with another researcher's result [9, 14]. Based on research results by Carrillo et al. [14], if the cellulose type I is more dominant than cellulose type II, the absorption at 1420 and 1155 cm^{-1} would shift to around 1430 and 1162 cm^{-1}, respectively. The presence of cellulose I is shown by the absorption at around 1430 and 1162 cm^{-1} in all sample spectra. Moreover, absorption at 893 cm^{-1} for cellulose type I would shift to around 897 cm^{-1}. The absorption for

Table 1 Peak assignments for absorption peak of each sample

Wavenumber (cm^{-1})	Peak assignment	Component	SF	SF A	SF A-T2h	SF A-T6h
3350	–OH str. intramolecular H-bonds	Cellulose I and II	3350	3350	3340	3340
2900	CH str	Cellulose I and II	2919	2917	2915	2917
1635	–OH of water absorbed cellulose	Cellulose I (1630), II (1620)	1636	1635	1634	1634
1420	CH_2 symmetric bend	Cellulose I (1430), II (1420)	1428	1428	1430	1431
1375	CH bend	Cellulose I and II	1372	1368	1367	1369
1315	CH_2 wag	Cellulose I (1317), II (1315)	1318	1316	1315	1315
1155	C–O–C asymmetric str	Cellulose I (1162), II (1155)	1160	1159	1159	1159
893	Group C_1 frequency	Cellulose I (897), II (893)	897	896	896	896

wag wagging, *str* stretching, *bend* bending

all samples is appeared at around 897 cm^{-1}. Additionally, absorption at 1315 cm^{-1} is identified for all fibers following treatment; these peaks verified the presence of cellulose type II on a modest scale. According to FTIR spectroscopy, cellulose I is the predominant molecule in sorghum both untreated and after alkalinization and TEMPO-mediated oxidation.

3.2 *Fiber Morphology*

Figure 2 shows the FE-SEM image of untreated sorghum fiber, alkalinized sorghum fiber from our previous work [8] and alkalinization/TEMPO-mediated oxidation sorghum fiber. According to the FE-SEM image, the fiber becomes defibrillated over the alkalinization and more defibrillated after TEMPO-mediated oxidation treatment. Figure 2a demonstrates that the untreated fiber's surface remains rough, with a width of around 400 m, comparable to that of a single bundle fiber, and the waxes as impurities remain masked the surfaces. Alkalinization treatment causes surface fibers to be more clear from impurities, and some parts of fibers begin defibrillated into single fibers as seen in Fig. 2b. Hemicellulose and lignin have been eliminated from fiber over alkalinization, which causes defibrillation fiber into a single bundle fiber [15]. Figure 2c illustrates the shape of the fiber surface following alkalinization and the TEMPO-mediated oxidation process. In this case, the fibers become cleaner and more fibrillated into microfibrils. As confirmed by the FTIR spectrum in Fig. 1,

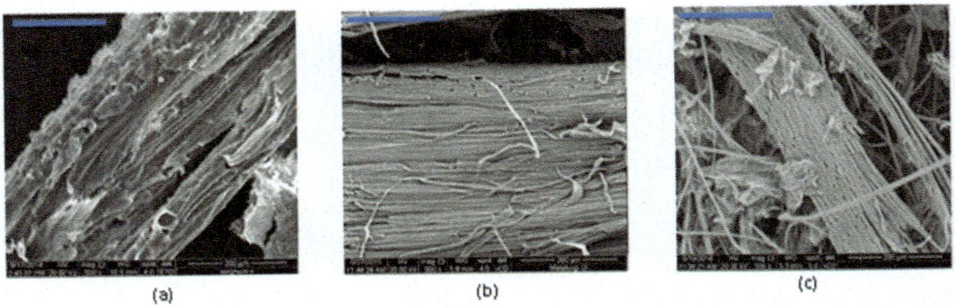

Fig. 2 Morphological of sorghum fibers, **a** untreated [8], **b** alkalinization [8], **c** alkalization/TEMPO-mediated oxidation (blue bar scale represents 200 μm)

the TEMPO-mediated oxidation has a significant effect on the elimination of hemicellulose and lignin. This morphology indicates the separation of single microfibrils with the small-scale fibers approximately 10 microns, but some fibers bundles remain after the treatment.

3.3 *Fiber Crystallinity*

In general, cellulose structures are made of amorphous and crystalline areas in varying proportions depending on the origin and treatment procedure, with hemicellulose and lignin having an amorphous structure. The XRD pattern of untreated and alkalinized sorghum fibers followed by TEMPO-mediated oxidation is shown in Fig. 3a. This treatment causes the sorghum fibers to shed part of their lignin and hemicellulose, while increasing the crystalline ratio. The XRD characterization result for sorghum fibers illustrates the enhancement of the crystalline areas. Untreated sorghum fibers

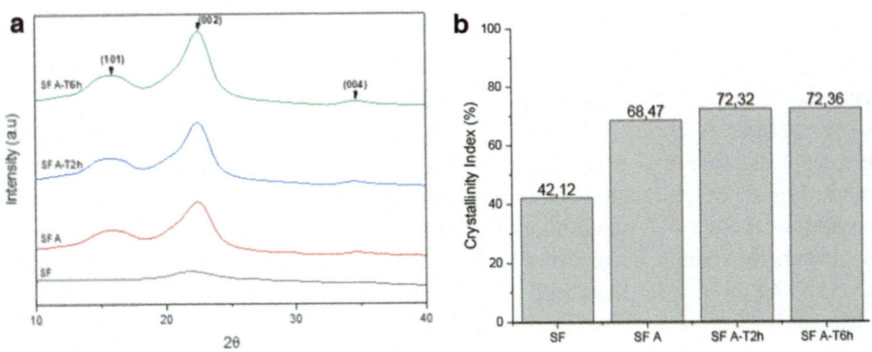

Fig. 3 XRD diffractograms **a** and their crystallinity index **b** of sorghum untreated and treated by alkalinization and TEMPO-mediated oxidation

have three significant peaks at around 16.0, 22.0 and 34.0°. These peaks correspond to the type I crystal structure of cellulose in the area diffractions (101), (002) and (040) [16].

The peak in the (002) plane grows sharper for sorghum fibers treated with alkalinization and alkalinization/TEMPO-mediated oxidation [17]. We infer that alkaline and alkaline/TEMPO oxidation treatments improve cellulose crystallinity by removing some lignin and hemicellulose. Thus, the cellulose content was increased, resulting in a stronger peak in the (002) plane. The maximum crystallinity of sorghum fibers was achieved using an alkalization treatment followed by an oxidation process driven by TEMPO. Figure 3b compares the crystallinity index obtained from both modification procedures to that obtained from untreated sorghum fibers. Equation (1) was used to obtain the value from XRD curves. As seen in Fig. 3b, the most successful method for removing the amorphous sections of sorghum fibers was alkalization followed by TEMPO-mediated oxidation. The crystallinity index of cellulose suggests a more compact cellulose structure that makes heat transmission more difficult, perhaps increasing their thermal stability [18].

4 Summary

The purpose of this work is to characterize MFC derived from sorghum bicolor utilizing a graduated alkalinization followed by TEMPO-mediated oxidation. After treatment, the FTIR spectrum revealed the absence of binding elements in fibers such as hemicellulose and lignin, as well as the predominant polymorph cellulose type I. The TEMPO-mediated oxidation has a considerable influence on the morphology of MFCs, implying that single microfibrils are separated by around 10 microns, while some fiber bundles remain. The XRD diffraction analysis revealed that alkalization treatment followed by TEMPO-mediated oxidation resulted in the greatest crystallinity index of 72 percent.

References

1. Roziafanto AN, Dwijaya MS, Yunita R, Amrullah M, Chalid M (2019) Synthesis hybrid biopolyurethane foam from biomass material. In: AIP conference proceedings vol 21751, no 1. AIP Publishing LLC, p 020068
2. Lavoine N, Desloges I, Dufresne A, Bras J (2012) Microfibrillated cellulose–Its barrier properties and applications in cellulosic materials: A review. Carbohyd Polym 90(2):735–764
3. Yuanita E, Pratama JN, Mustafa JH, Chalid M (2015) Multistages preparation for microfibrillated celluloses based on Arenga pinnata "ijuk" fiber. Procedia Chemistry 16:608–615
4. Christwardana M, Handayani AS, Savetlana S, Lumingkewas RH, Chalid M (2020) Microfibrillated cellulose fabrication from empty fruit bunches of oil palm. In: Materials science forum, vol 1000. Trans Tech Publ, pp 272–277

5. Husnil YA, Yuanita E, Ramadhani N, Chalid M (2020) Study on the effect of bleaching treatment on the mechanical properties of kenaf fibers. In Materials science forum, vol 1000. Trans Tech Publ, pp 278–284
6. Novovic A, Lazwardi DR, Zulfia A, Chalid M (2019) Microfibrillated cellulose (MFC) isolation based on stalk sweet sorghum through alkalinization-bleaching treatment: effect of soaking temperature. IOP Conf Ser Mater Sci Eng 509(1):012079
7. Fatriasari W, Iswanto AH (2015) The kraft pulp and paper properties of sweet sorghum bagasse (Sorghum bicolor L Moench). J Eng Technol Sci 47(2)
8. Ismojo I, Ammar AA, Ramahdita G, Zulfia A, Chalid M (2018) Influence of chemical treatments sequence on morphology and crystallinity of sorghum fibers. Indonesian J Chem 18(2):349–353
9. Handayani S, Husnil YA, Handayani AS, Chalid M (2019) Application of waste sorghum stem (sorghum bicolour) as a raw material for microfibre cellulose. IOP Conf Ser Mater Sci Eng 509(1):012015
10. Tanaka R, Saito T, Isogai A (2012) Cellulose nanofibrils prepared from softwood cellulose by TEMPO/NaClO/NaClO2 systems in water at pH 4.8 or 6.8. Int J Biol Macromol 51(3):228–234
11. Segal L, Creely JJ, Martin A Jr, Conrad C (1959) An empirical method for estimating the degree of crystallinity of native cellulose using the X-ray diffractometer. Text Res J 29(10):786–794
12. Roziafanto AN, Alfarisi F, Ramadhan T, Chalid M (2020) Preliminary study of modified lignin compatibility in polypropylene-modified bitumen. Macromol Symp 391(1):1900158
13. Célino A, Gonçalves O, Jacquemin F, Fréour S (2014) Qualitative and quantitative assessment of water sorption in natural fibres using ATR-FTIR spectroscopy. Carbohyd Polym 101:163–170
14. Carrillo F, Colom X, Sunol J, Saurina J (2004) Structural FTIR analysis and thermal characterisation of lyocell and viscose-type fibres. Eur Polymer J 40(9):2229–2234
15. Kalia S, Kaith B, Kaur I (2009) Pretreatments of natural fibers and their application as reinforcing material in polymer composites—a review. Polym Eng Sci 49(7):1253–1272
16. El Oudiani A, Chaabouni Y, Msahli S, Sakli F (2011) Crystal transition from cellulose I to cellulose II in NaOH treated Agave americana L. fibre. Carbohyd Polym 86(3):1221–1229
17. Kobayashi K, Kimura S, Togawa E, Wada M (2011) Crystal transition from Na–cellulose IV to cellulose II monitored using synchrotron X-ray diffraction. Carbohyd Polym 83(2):483–488
18. Poletto M, Ornaghi HL, Zattera AJ (2014) Native cellulose: structure, characterization and thermal properties. Materials 7(9):6105–6119

A Self-Healing Study of Polymeric Films Made from Carboxylated Nitrile Butadiene Rubber Latex-Containing Reactive Epoxidized Crosslinker

Nuratiqah Ab Samad, Rohah A. Majid, ZhenLi Wei, and YiFan Goh

1 Introduction

Nitrile rubber (NBR) is a special-purpose rubber comprised of about 12% of global rubber consumption [1]. It is used in a broad range of applications depending on the nitrile polymer designs which result in specific characteristics that suit its final applications. As an example, carboxylated nitrile butadiene rubber (XNBR) has been widely applied in the rubber glove industry as an alternative to natural rubber (NR). XNBR designed with functionalized pendant carboxylic acid leads to rubber with improved strength and resistance toward chemical and abrasion. These improvements are achieved through vulcanization which induces the formation of a bonding network by the active functional groups in the designed polymer that may comprise of covalent and/or non-covalent network.

Nevertheless, it was realized that the current fundamental of polymer engineering adapted in elastomers is constructed by irreversible covalent bonds which give disadvantages in terms of recovery of the polymer [2]. When deformation of the polymer occurs, the non-covalent bond normally dissociates first due to lower bond energies and can reform upon removal of the force. However, in the case of the covalent bond, after removal of the deformation, the bond breaks permanently [3, 4]. Due to this fact, the final shape of the polymer remains, consequently restricting the polymer from being reprocessed which is exactly the case of vulcanized XNBR.

N. A. Samad · R. A. Majid (✉)
School of Chemical and Energy Engineering, Faculty of Engineering, Universiti Teknologi Malaysia, 81310 Johor Bahru, Johor, Malaysia
e-mail: r-rohah@utm.my

N. A. Samad · Z. Wei · Y. Goh
Synthomer Sdn Bhd, No. 73, Jalan i-Park 1/8, Kawasan Perindustrian i-Park, Bandar Indahpura, 81000 Kulaijaya, Johor Darul Takzim, Malaysia

© The Author(s), under exclusive license to Springer Nature Singapore Pte Ltd. 2023

S.-M. Chen (ed.), *Proceedings of 10th International Conference on Chemical Science and Engineering*, Springer Proceedings in Materials 21,
https://doi.org/10.1007/978-981-19-4290-7_3

Zhang et al. [2] comprehensively reviewed the techniques that can be applied in developing polymers with reversible covalent chemistry which enable the polymers to be intrinsically self-healable. These techniques are categorized into general two-step fashion and dynamic reversible covalent reaction (one-step fashion). Concerning XNBR, the dynamic reversible covalent reaction is considered as a practical approach where the presence of carboxylic acid in the polymer network can be paired with potential curing additives that can impart the XNBR with intrinsic self-healing property. However, there have been limited works demonstrating the possibility of dynamic reversible covalent reaction in XNBR, whereby most of the works focused on developing fabricated NBR with non-covalent reactions and metal coordination bonds which are well known for their thermal-reversible characteristics [3, 5–7].

Since the revelation of the novel discovery of vitrimers by Liebler et al. (2011), few interesting works have successfully demonstrated reversible covalent chemistry by transesterification through epoxy-acid reaction. For instance, bio-based epoxidized soy-bean oil (ESO) was either used as curing additives [8] or as a base polymer cured by polycarboxylic acids [9]. Both studies managed to prove that the dynamic rearrangement of β-hydroxyester linkages generated from the epoxy-acid reaction can participate in transesterification reactions at the appropriate temperature, assisted by a catalyst, resulting in effective self-healing of the polymer. In a broad sense, implementing epoxy-acid reaction in XNBR is reckoned as one of the possible strategies for self-healed and even recycling of the materials. Besides, the commercial preparation methods of carboxyl- and epoxy-functionalized elastomers are favorable for large-scale production which advantageous for practical use [8].

In conjunction with sustainable rubber development, the introduction of self-healing property into commercial rubber is essential to produce high-end elastomeric products with an extended lifetime. Therefore, the objective of this study is to develop XNBR with self-healing capability enabled by an epoxy-acid reversible crosslink network and to use an external heat stimulus to trigger the bonds' breaking and reforming. As known, a considerable amount of catalyst is necessary for transesterification reaction to achieve polymers with excellent recyclability and robust properties [2]. Herein, the implication of employing a common ZnO catalyst used to accelerate the crosslinking reaction in XNBR upon the transesterification reaction of the XNBR was investigated. It is also worth noting that the reaction between XNBR and ZnO generates an ionic crosslinks network which helps in the enhancement of the XNBR physical and mechanical properties. Since XNBR combines both covalent and non-covalent crosslinking networks, this study focused on the effects of the combination on the physical and the tensile properties of XNBR, as well as the influence of pH adjustment on the healing efficiency of XNBR.

2 Experimental

2.1 *Synthesis of Nitrile Polymer/XNBR Through Emulsion Polymerization*

The emulsions polymerizations were carried out in a nitrogen-purged autoclave. The reducing agent (Bruggolite FF6) and sodium persulfate were first added to the autoclave and subsequently heated to 30 °C. Then, the monomers (acrylonitrile, butadiene, and methacrylic acid) were added into the autoclave together with tertiary dodecyl mercaptan (TDM). Over 10 h, sodium dodecylbenzene sulfonate solution was added. The temperature was maintained at 30 °C up to a conversion of 95%, resulting in a Total Solids Content (TSC) of 45%. The polymerization was short-stopped by the addition of diethyl hydroxylamine. The pH was adjusted using potassium hydroxide (5% aqueous solution) to pH 7.5, and the residual monomers were removed by vacuum distillation at 60 °C. Then, the pH was adjusted to 8.2 by the addition of potassium hydroxide solution.

2.2 *Synthesis of Reactive Epoxidized Crosslinker (REC)*

Initially, a nitrogen-purged autoclave was charged with diphenyl oxide disulfonate dissolved in water heated to a temperature of 70 °C. TDM was added to the initial charge, together with ammonium peroxodisulfate solution. Then, the monomers (butadiene, acrylonitrile and glycidyl methacrylate or GMA) and a solution of diphenyl oxide disulfonate dissolved in water were added. After the addition of the monomers, the temperature was maintained at 70 °C. The polymerization was maintained up to a conversion of 99%. The TSC of the REC is 38%, and its pH is 7.

2.3 *Preparation of Polymer Blending of XNBR with REC*

A series of polymer blends comprising of XNBR and different REC contents was prepared as presented in Table 1. XNBR was blended with REC (0, 10, 25 wt%) and stirred for at least 2 h to ensure homogeneity.

Preparation of XNBR/REC rubber composites/films: The resulting polymer blend was compounded with additives based on part per hundred rubber (phr) at room temperature and according to the formulations shown in Table 2. First, the polymer blend was added with the additives and constantly stirred for at least 16 h to achieve a maturation period. Next, vulcanization was done by casting the compounds on a glass plate and allowed to dry at room temperature about 25 °C for 3 days. The obtained films were then annealed in an oven at 90 °C for 24 h to ensure complete

Table 1 Polymer blend recipes

Samples	Weight of XNBR emulsion with TSC of 45% (g)	Weight of REC with TSC of 38% (g)	Wt% of GMA in the blend (part per monomer)
XNBR-100/REC-0	100	0	0
XNBR-90/REC-10	90	10	3
XNBR-75/REC-25	75	25	8.5

Table 2 Rubber compound formulations

Samples	Sulfur (S) (phr)	Zinc diethyldithiocarbamate (ZDEC) (phr)	Zinc oxide (ZnO) (phr)
XNBR-100/REC-0	0.8	0.7	1
XNBR-90/REC-10	0 0	0 0	0 1
XNBR-75/REC-25	0 0	0 0	0 1

drying and crosslink formation. Then, vulcanized specimens were cut into certain shapes and dimensions needed by the characterization techniques.

Another series of XNBR/REC films were prepared at different pH values (0, 5, 10, 11, and 11.5) using samples with 10% REC content without the presence of ZnO.

2.4 Characterization

Fluid properties: Total solids content (TSC), pH value, z-average particle size and viscosity were used to analyze the fluid properties of the XNBR/REC latex. The TSC test was based on a gravimetric method. Approximately, 1 g (m_{initial}) of the latex sample was weighed on an analytical balance into a tarred aluminum dish and heated at 120 °C for 1 h in a circulating air oven until the constant mass was reached. After cooling to room temperature, the final weight (m_{final}) was measured. The TSC was calculated according to Eq. (1):

$$\text{TSC} = \frac{m_{\text{initial}}}{m_{\text{final}}} \times 100\% \tag{1}$$

The pH value was determined according to DIN ISO 976 using a Schott CG 840 pH meter electrode calibrated with buffer solutions of pH 7 and pH 10.01. The electrode was immersed in the latex dispersion at room temperature of about 25 °C, and the constant value on the display was recorded as pH value.

The *z*-average particle size was measured based on the dynamic light scattering method using a Malvern Zetasizer Nano S (ZEN 1600). The latex was transferred into a test cuvette and diluted using deionized water to a certain turbidity level. The cuvette was gently mixed to produce homogenized samples which were then placed in the measurement device where the value obtained was recorded as *z*-average particle size.

The viscosity measurement was carried out using a Brookfield LVT viscometer. Approximately, 220 ml of the latex was filled into a 250 ml beaker and the spindle number 1/60 was used and immersed up to the mark. The viscometer was run for 1 min, and the constant value was recorded.

Tensile Properties: The dried films were cut into 3 mm type D dumbbell shape specimens and tested according to the ASTM D412-06a standard using a Universal Testing Machine (Zwick Roell Z005 TN Proline). The test was carried out at room temperature and the relative humidity of 50 ± 5% using a crosshead speed of 500 mm/min. An average of at least five specimens was taken for each experimental point to obtain reliable values.

Healing efficiency: The film was cut into halves before manually repositioned together and pressed for 60 s. Then, the pressed film was held with a wooden peg to ensure full contact (see Fig. 1). The film was healed by heat treatment in an oven at 180 °C for 30 min. The healed sample was subjected to a stress–strain test following the same tensile conditions. The healing efficiency (η) was calculated as a measure of the retention on tensile strength, as follows:

$$\eta(\%) = \frac{P_{\text{Healed}}}{P_{\text{Pristine}}} \times 100 \quad (2)$$

where P_{Pristine} and P_{Pristine} are the values of original and healed of the sample tensile strength, respectively.

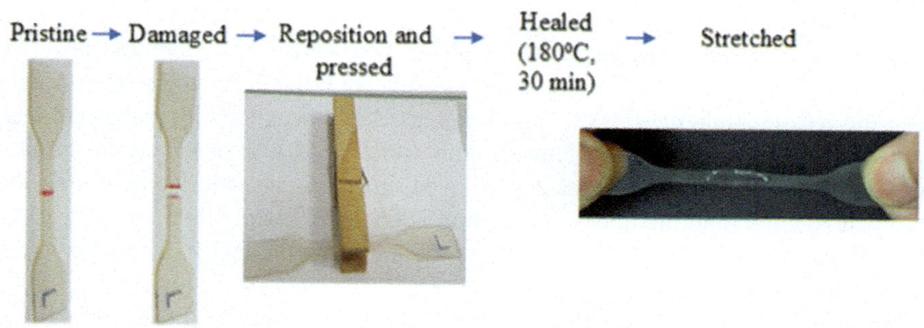

Fig. 1 Representative pictures of the cut-and-heal process

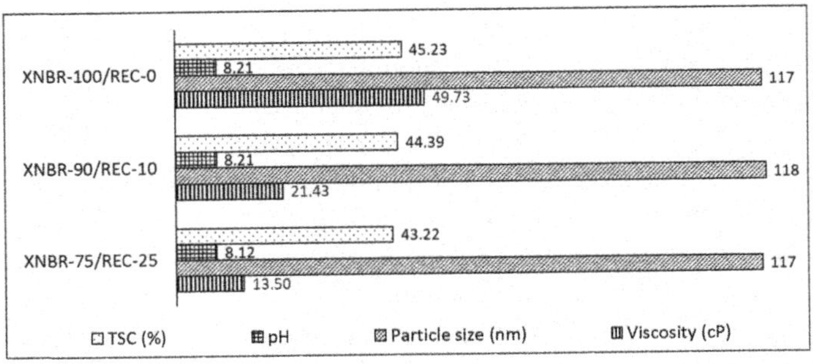

Fig. 2 Fluid properties of XNBR/REC blends

3 Results and Discussions

3.1 XNBR/REC Blends

Figure 2 shows the fluid properties of the pristine XNBR latex and XNBR/REC blends at different ratios. The pH and particle size of all samples were similar, i.e., 8.1–8.2 and 117 nm, respectively. By increasing the amount of REC had decreased the viscosities of the blends due to lower TSC.

3.2 Effect of REC and ZnO Content on the Physical Properties and Healing Capability of XNBR/REC Compounds

The physical properties of the pristine XNBR and XNBR/REC blend at different ratios were shown in Table 3. For samples with ZnO, the ultimate tensile strength and the tensile strength at 100% strain increased, whereas the elongation at break decreased with higher REC contents. Similar trends were also observed for samples without ZnO. It was found that ZnO played an important role in increasing the ultimate strength and the tensile strength at 100%, 300% and 500% strain, as seen in XNBR-90/REC-10 blend. This reflects the influence of ZnO in determining the final properties of nitrile gloves [10].

In *terms* of healing capability, XNBR-90/REC-10 without ZnO exhibited the most promising performance where the recovery percentages in elongation at break reached >95% and the tensile strength was >70% as manifested in Fig. 3. It seems that the presence of ZnO inhibited the healing capability. It was thought that the ZnO was competing with the epoxy-functional groups in the REC during the reaction with carboxylic acid groups of the methacrylic acid in XNBR latex. This reduced the availability of β-hydroxyester linkage in the XNBR which was crucial for healing capability. It was also found that the incorporation of excessive epoxy groups into

Table 3 Physical properties of XNBR/REC compounds

Compound	XNBR-100/REC-0	XNBR-90/REC-10	XNBR-90/REC-10	XNBR-75/REC-25	XNBR-75/REC-25
S (phr)	0.8	0	0	0	0
ZDEC (phr)	0.7	0	0	0	0
ZnO (phr)	1	0	1	0	1
pH	10	10	10	10	10
M_{100}	1.1	1.1	1.5	1.9	2.5
M_{300}	1.8	1.8	3.4	-	-
M_{500}	3.4	3.5	10.9	-	-
Elongation at break (%)	575	672	588	258	285
Tensile strength (MPa)	5.3	7.7	19.7	12.3	22.3
Recovery percent of tensile strength (%)	43	73	20	14	13
Recovery percent of elongation at break (%)	62	96	60	33	49

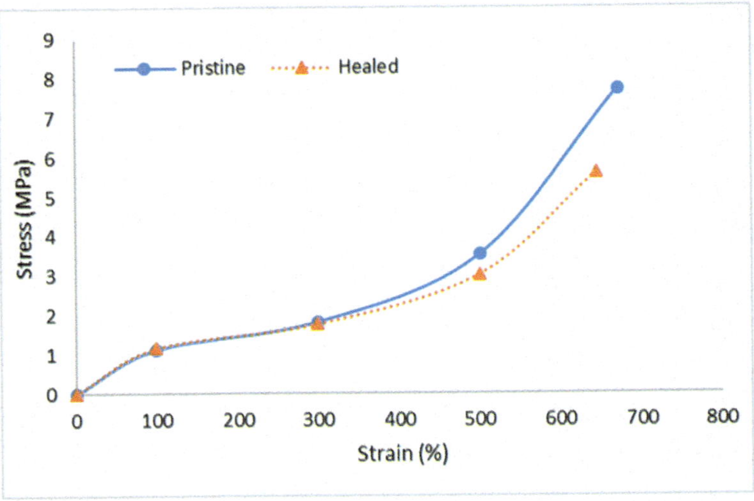

Fig. 3 Stress–strain curve of XNBR-90/REC-10 without ZnO

the system (XNBR-75/REC-25) had reduced the healing capability of the system. It was thought that some excessive epoxy groups have been hydrolyzed at high pH (in this case pH 10) [11] and reacted with other epoxy groups to form irreversible bonds. Hence, the healing capability decreased significantly with the formation of such irreversible bonds. Another possibility could be related to higher amounts of crosslinking which restricted the movement of the polymer chains, causing a difficulty in the bond rearrangement, thus reducing the healing capability [12].

3.3 *Effect of pH Value on the Physical Properties and Healing Capability of XNBR/REC Compounds*

The ionization state of carboxylic acid groups in the XNBR latex depends on the pH value. For example, at pH 5, most of the carboxylic acid groups are in the form –COOH, whereas at ~pH 10, at least half of the carboxylic acid groups are in the form of $-COO^-$. Hence, to get more understanding, a blend of XNBR-90/REC-10 without ZnO was selected as a model to study the effect of pH value on the physical properties and healing capability, as the blend exhibited the best elongation at break and healing capability as previously discussed.

From Table 4, it was found that at pH 5, the tensile strengths at 100%, 300% and 500% strains exhibited the lowest values among all samples but the elongation at breaks were the highest, i.e., >1400%. The reasons could be due to most of the carboxylic acid groups are in the $-COOH$ form and there was no ionic crosslinking occurred between $-COO^-$ and K^+ from potassium hydroxide (i.e., pH adjuster). When the pH was increased to 10 and 11.5, the number of $-COO^-$ groups was also

Table 4 Effect of pH value on physical properties XNBR/REC compounds

Compound	XNBR-100/REC-0	XNBR-90/REC-10	XNBR-90/REC-10	XNBR-90/REC-10
S (phr)	0.8	0	0	0
ZDEC (phr)	0.7	0	0	0
ZnO (phr)	1	0	0	0
pH	10	5	10	11.5
M_{100}	1.1	1.0	1.1	1.1
M_{300}	1.8	1.6	1.8	2.2
M_{500}	3.4	2.1	3.5	5.6
Elongation at break (%)	575	1424	672	573
Tensile strength (MPa)	5.3	5.6	7.7	8.3
Recovery percent of tensile strength (%)	43	50	73	95
Recovery percent of elongation at break (%)	62	36	96	81

increased, enhancing the level of ionic crosslinking. The arguments were supported by the increase of the ultimate tensile strength and the tensile strength as well as the decrease of elongation at break at 300% and 500% strains, respectively.

In terms of healing capability, XNBR-90/REC-10 at pH 10 showed the most promising performance. It seems that the healing capability is better for samples with higher alkaline pH, i.e., pH 10 and pH 11.5. At pH 5, the recovery of physical properties is not satisfactory which may be due to less formation of β-hydroxyester because of less epoxy-carboxylic acid reaction under this condition.

4 Conclusion

An epoxy polymeric crosslinker was used to induce β-hydroxyester crosslinking in carboxylated nitrile butadiene rubber latex, to impart self-healing capability to the rubber system. It was demonstrated that by incorporating 10% of the crosslinker, a good self-healing capability with the efficiency of >95% recovery of elongation at break and >70% recovery of tensile strength were achieved. Further discovery revealed that pH value plays a key role in promoting the self-healing capability, with alkaline pH giving the best performance. In conclusion, this study indicates the possibility of developing nitrile gloves with unique self-healing and recyclability features that may be beneficial for the rubber glove industry.

References

1. Imbernon L, Norvez S (2016) From landfilling to vitrimer chemistry in rubber life cycle Eur. Polym J 82:347–376
2. Zhang ZP, Rong MZ, Zhang MQ (2018) Polymer engineering based on reversible covalent chemistry: a promising innovative pathway towards new materials and new functionalities prog. Polym Sci 80:39–93
3. Utrera-Barrios S, Araujo-Morera J, Pulido de Los Reyes L, Manzanares RV, Verdejo R, LopezManchado MA, Santana MH (2020) An effective and sustainable approach of achieving self-healing in nitrile rubber. Eur Polym J 139
4. Sordo F, Mougnier S-J, Loureiro N, Tournilhac F, Michaud V (2015) Design of self-healing supramolecular rubbers with a tunable number of chemical cross-links. Macromolecules 48(13):4394–4402
5. Lipinska M, Gaca M, Zaborski M (2020) Curing kinetics and ionic interactions in layered double hydroxides-nitrile rubber M-Al-LDHs-XNBR composites Polym Bull
6. Cheng Z, Yan M, Cao L, Huang J, Yuan D, Chen Y (2019) Design of nitrile rubber with high strength and recycling ability based on Fe3+-catechol groups coordination ind. Eng Chem Res 58(9):3912–3920
7. Liu W-X, Zhang C, Zhang H, Zhao N, Yu Z-X, Xu J (2017) Oxime-based and catalyst-free dynamic covalent polyurethanes. J Am Chem Soc 139(25):8678–8684
8. Zhang G, Liang K, Feng H, Pang J, Liu N, Li X, Zhou X, Wang R, Zhang L (2020) Design of epoxy-functionalized sbr with bio-based dicarboxylic acid as cross-linker toward green curing process and recyclable ability. Ind Eng Chem Res 59(22):10447–10456
9. Altuna FI, Pettarina V, Williams RJJ (2013) Self-healable polymer networks based on the cross-linking of epoxidized soybean oil by an aqueous citric acid solution. Green Chem 15:3360–3366
10. Tan KY, Phang SW, Phang CK, Choh JL (2018) Preliminary study on effect of chemical composition alteration on elastic recovery and stress recovery of nitrile gloves. MATEC Web Conf 152(01011):172–183
11. Bonollo S, Lanari D, Vaccaro L (2011) 2011 Ring-Opening of Epoxides in Water. Eur J Org Chem 14:2587–2598
12. Irigoyen M, Matxain JM, Ruiperez F (2019) Effect of molecular structure in the chain mobility of dichalcogenide-based polymers with self-healing capacity. Polymers (Basel) 11(12):1960

Feasibility Study of Latex Stability for Free Solvent Hydrogenation to Natural Rubbers

Dody Andi Winarto, Chandra Liza, Awang Pemuji, and Mochamad Chalid

1 Introduction

Natural rubber (NR) latex is milky sap produced by *havea braziliensis* plant. It is colloidal system of rubber particles with diameter 0.1–3 micron and non-rubber particles suspended in aqueous medium. Rubber particles in NR latex compose of cis-1,4-poly isoprene. Fresh latex contains 30–40% natural rubber particles [1] and preserved in concentrated condition using ammonia. Concentrated latex is then packed and transported as raw material for further process to make various rubber products such as gloves, mattress, tires, shoes, seal, and automobile parts. This wide range applications is due to NR material has good elasticity, good processing characteristics, excellent physical and mechanical properties. In marine application, it is widely used as buoy/floater of fish net because of its superiority [2]: lightweight, good buoyancy and impact damping. Even this kind of material can be used as replacement of steel as buoyance material [3].

However, NR has some weaknesses. It has poor heat-, UV-, oxidation- and ozone-resistance. Hence, NR will be more brittle and their hardness will increase faster when exposed to the environment including sunlight. This is because NR contains unsaturated carbon double bonds more than 98% in its molecular structure [4]. Effort was needed to overcome the drawback by chemical modification, copolymerization or blending. Chemical modification can be performed by alteration the backbone of the structure such as hydrogenation, halogenation or epoxidation [5]. Among

D. A. Winarto · A. Pemuji · M. Chalid (✉)
Department of Metallurgical and Materials, Faculty of Engineering, University of Indonesia, Kampus UI Depok, Depok 16424, West Java, Indonesia
e-mail: chalid@metal.ui.ac.id

D. A. Winarto · C. Liza
Balai Teknologi Polimer, Badan Riset Dan Inovasi Nasional, Gedung #460, Puspiptek Area, Tangerang Selatan 15314, Banten, Indonesia

© The Author(s), under exclusive license to Springer Nature Singapore Pte Ltd. 2023
S.-M. Chen (ed.), *Proceedings of 10th International Conference on Chemical Science and Engineering*, Springer Proceedings in Materials 21,
https://doi.org/10.1007/978-981-19-4290-7_4

them, hydrogenation was an easier technique to implement. Catalytic hydrogenation can be done using precious transition metal, high-pressure reactor, organic solvent and high-temperature condition. Non-catalytic hydrogenation can be done using milder condition in atmospheric glass reactor, but still using organic solvent and high-temperature condition. Hydrazine hydrate and hydrogen peroxide system can be used as hydrogen resources to hydrogenate NR with water as solvent. This approach is considered green and reducing environmental risk [6, 7]. These hydrogenation were conducted in latex form, and the success rate of the reaction depends on the stability of the latex. Hydrogenation of NR latex using various surfactant and coagulation agent was reported [8]. NaOH was also used for as stabilizing agent in hydrogenation [6]. The stability also affected by temperature, rubber content and other [9]. The dispersity of latex can be indicated by the turbidity of the system [10].

In this paper, we report the screening of surfactant use to enhance the stability of the natural rubber latex system for preparing the hydrogenation using hydrazine hydrate and hydrogen peroxide as hydrogen source system. Furthermore, the experimental was performed to optimize the stability of the system through varying concentration of rubber, agitation rate, temperature and time. This study was evaluated by measuring turbidity and viscosity to analyze dispersion and distribution of rubber particles and rubber particles' resistance to agitation, respectively.

2 Material and Experimental Method

2.1 Materials

Fifty-six percentage dry rubber content (DRC) commercially high-ammoniated deproteinized concentrated natural rubber (NR) latex was provided by Indonesian Rubber Research Institute (IRRI, Bogor, Indonesia). Three surfactants: EMAL (anionic), SDS (anionic) and EMULGEN (nonionic) were obtained from PT KAO Indonesia.

2.2 Experimental

NR latex was diluted with aquades into 250 mL beaker glass to get 20% rubber concentration. The solution was then treated to attain pH 10. 1 phr surfactant was added into the beaker to stabilize the colloidal system. The solution was then stirred, using mechanical stirrer at 200 rpm. Temperature of the sample was regulated using water as heating media. The sample in 250 mL beaker glass was put into 1000 mL beaker glass filled with water which was placed on hot plate. The treatment was run for 5 h. After 1 h, dynamic viscosity was measured using Viscometer Thermo Haake 6 Plus. Beside that 10 mL sample was taken into sample bottle to measure turbidity

Table 1 Experimental condition

Variable	Base line	Range
Surfactant type	SDS	EMAL, SDS, EMULGEN
rubber content (%w/w)	20	20, 25, 30
Temperature (°C)	25	25, 50, 70
Agitation rate (rpm)	200	60, 100, 200
Time (hours)	3	1, 3, 5

of the sample using turbidity meter Lutron TU-2016. Consecutively every 2 h, the viscosity and turbidity were measured. Experiment were done according to process condition as stated in Table 1.

As an example of this experimental condition: Variation of rubber content with range of 20%, 25% and 30% experiment was conducted using SDS as surfactant, at 25 °C, agitation rate of 200 rpm and recording time of turbidity and viscosity at 3 h (Base line). The exception was that we use 56% rubber content for variation of agitation rate. Displayed equations are centered and set on a separate line.

3 Results and Discussion

Colloidal stabilization was an important requirement in preservation NR latex before processed. It is to anticipate the environmental condition such as oxidation, heat, mechanical, bacterial, for the latex to undergo further processing. If the latex was not stable, flocculation could occur and NR latex will be difficult to process. This stabilization is due to the maintained repellation force between rubber particles. Protein membrane with negative charge was enveloping the natural rubber polymer [11] as seen in Fig. 1. The stability is also mean that the particles were scattered randomly. Mechanical agitation could help particles and heat to be distributed in

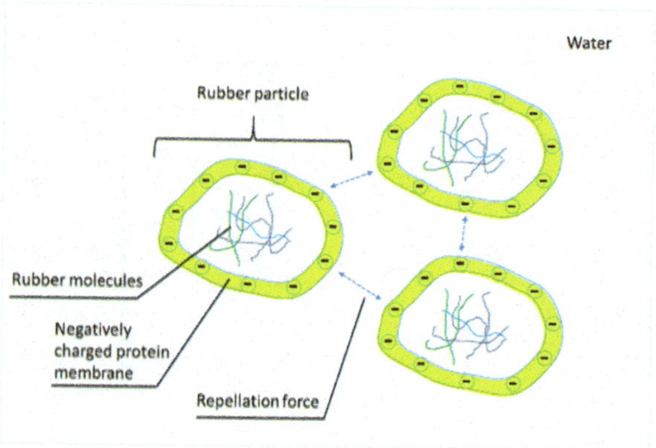

Fig. 1 Proposed schematic rubber particles in water [11]

the system. This stability can be recognize by measuring turbidity of the latex. If flocculation occurred, the turbidity will decrease [12]. On the other hand, viscosity was supporting indicator to compare the resistance of solution to mechanical or electrical disturbance.

3.1 Screening of Surfactants

In these experiment, two types of surfactant were used, anionic (EMAL and SDS) and nonionic surfactant (EMULGEN, see Table 1). Processes were carried out at temperature of 25 °C, agitation rate of 200 rpm and rubber concentration 15% (w/w). Viscosity of colloidal system as a function of time was measured, and the value was constant during the measurement time. Viscosity was stable in the system using any surfactant. However, colloidal system using SDS has higher viscosity value (13 mPa.s) which was 8.3% higher compared to EMAL and EMULGEN. System with high viscosity indicated stronger interaction between particles [13]. Figure 2 shows turbidity of colloidal system as function of time. The smaller the change of turbidity during experimental time indicates the stability of the system. The change of turbidity of the system using SDS was relatively smaller than of the system using EMAL and EMULGEN. The changes were 6.2% and 25.7% every two hours. On the other hand, turbidity changes of system using of EMAL and EMULGEN reached 78.9% and 102.6%. Stability mechanism of colloid system is due to electrostatic, steric and mixed mechanism between these electrostatic and steric mechanisms [10]. Even EMAL and SDS were the same type of anionic surfactant, refer to Fig. 2, and each turbidity profile was different. Hence, the stability of colloidal system was different. System with SDS was more stable than with EMAL. Therefore, there was a possibility that the mechanism of stability is not only due to electrostatic

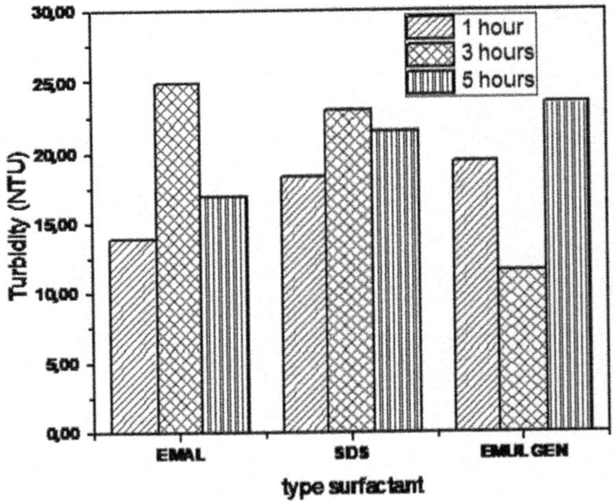

Fig. 2 Turbidity of solution using different types of surfactant

mechanism, but mixed one. From the above viscosity and turbidity data, SDS was chosen as surfactant for further optimization process.

3.2 Natural Rubber Compositions

The latex system will be used for hydrogenation reaction using NR latex. Water was used as solvent and NR latex as raw material. Some important parameters for the reaction includes number of particles involves (natural rubber latex concentration), temperature and necessity of agitation. Turbidity of the system increased as concentration (rubber content, %w/w) increased as shown in Fig. 3. Increasing concentration means there were more particles of rubber in the system. The particles theoretically were moving through Brownian movement [13]. As the amount of particles increased, the probability of turbidity value was also increased. This is due to more particles spread out in the solution, and it will reduced the scattered of incident light. Figure 3 shows that the viscosity increases as concentration (rubber content, from 20 to 30%) increases. As concentration was increased, more particles in the system was also increased. Interaction between particles due to repellation force between particles coupled with surfactant addition effect required higher force to agitate the system.

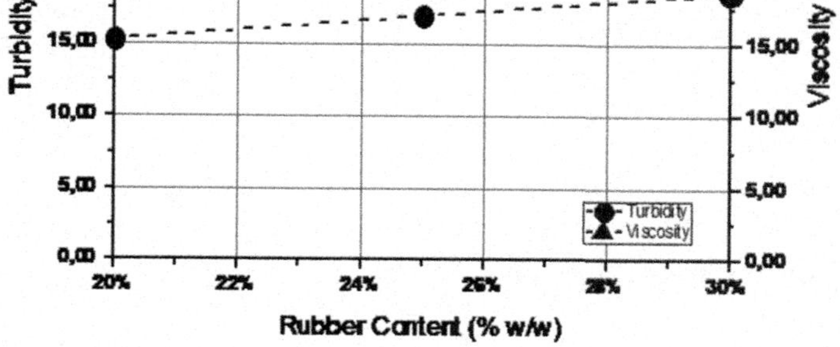

Fig. 3 Turbidity and viscosity of solution in different rubber compositions

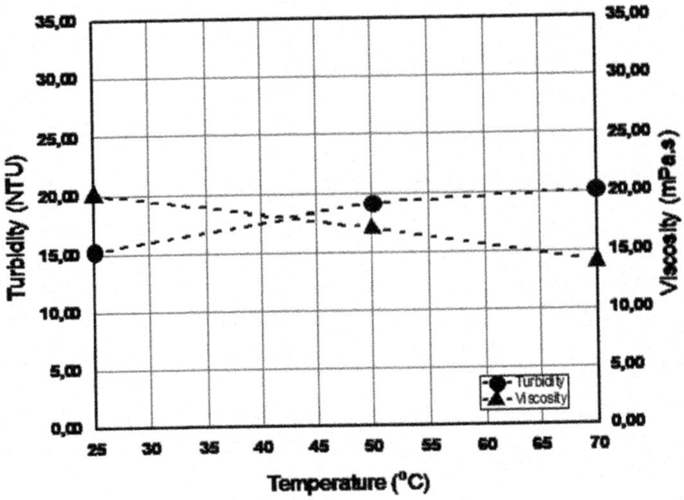

Fig. 4 Turbidity and viscosity of solution in different temperatures

3.3 *Temperatures*

As previously reported [9], there are some mechanical treatment which effected suspended particles in fluid, for example agitation and heating condition. Increasing temperature induced Brownian movement more active. Hence, the probability of NR particles to collide was increased. Figure 4 shows that turbidity of latex colloid increased from 25 °C and still increases until 70 °C although the slope of rising of the turbidity smaller than the increase from 25 °C to 50 °C. At higher temperature, turbidity decreased. Initial flocculation may have been occurred. It is because with constant agitation rate and increasing temperature, membrane of particles could be damaged and flocculation grown. This can be observed in decreasing viscosity. The condition of the process at 90 °C can be seen in Fig. 5. It is seen that coagulation was occurred clearly.

3.4 *Agitation Rates and Time*

Agitation is a mean, made for the system to accelerate the homogenization when new reactants introduced into it. Agitations were varied by agitation rate and time of agitation. The change of turbidity and viscosity was observed.

Agitation rates. Figure 6 shows the turbidity and viscosity of the system as a function of the magnitude of agitation. Agitation rate was varied between 60 and 200 rpm. As rpm increased, turbidity value was also increased. Agitation escalates collision of particles. It also gives the system mechanical energy that eventually increased the

Fig. 5 NR latex condition 200 rpm at 90 °C 4–5 h

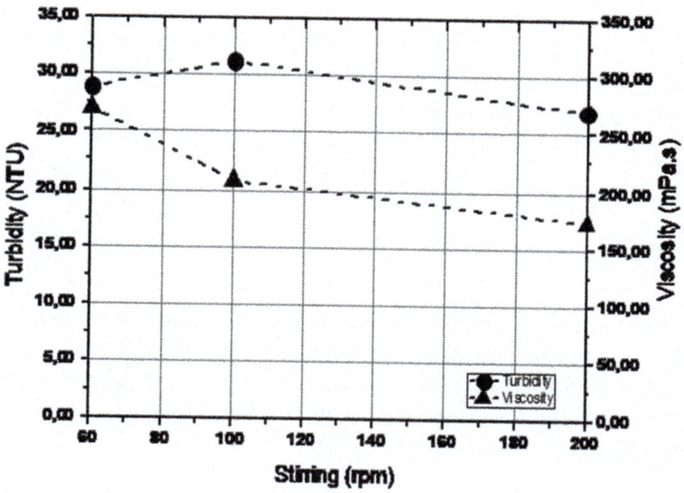

Fig. 6 Turbidity and viscosity of NR latex solution in different agitation conditions

temperature. Increasing temperature will increase Brownian movement and probability of flocculation. Turbidity value has a peak at 100 rpm, and decreasing at higher rpm. It is considered that the flocculation was observed at 100 rpm. Figure 6 also shows viscosity was decreased as increasing agitation rate rpm. This is an indicator that the system was non-Newtonian fluid and classified as a shear-thinning type. Micro-structural rearrangement [14] was occurred as the shear stress is increased due to increasing of agitation rate rpm. And, this expands the possibility of damaging the membrane which can be seen as reduction in turbidity value.

Agitation time. Figure 7 shows the turbidity of the system as a function of time. It was collected at 1, 3 and 5 h. The condition of the system were rubber content

Fig. 7 Turbidity of NR latex solution as function of time

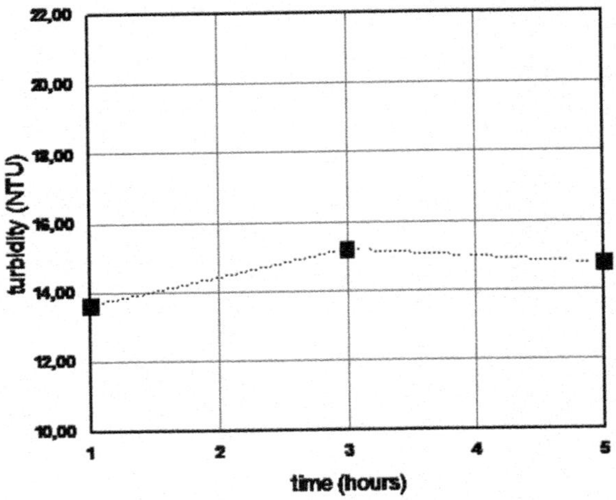

20% (w/w) and agitation rate 200 rpm. As shown in the figure, there exist peak on turbidity at reaction time of 3 h. It is indicated that the particles grown bigger and tend to move down as turbidity decreased at 5 h.

4 Summary

Viscosity and turbidity measurement can be used to assess the stability of colloid system of NR latex for preparing hydrogenation of NR latex. Within our experimental condition showed in Table 1, if SDS 1%w/w as surfactant was used, the optimum stability of latex was obtained at 20%w/w of rubber concentration, 100 rpm of agitation rate during 3 h at 25 °C.

References

1. Lim H, Misni M (2016) Colloidal and rheological properties of natural rubber latex concentrate. Appl Rheol 26(1):25–34
2. Lee E-K, Choi S-Y (2007) Preparation and characterization of natural rubber foams: effects of foaming temperature and carbon black content. Korean J Chem Eng 24(6):1070–1075
3. Park YW (2016) Design of Korean standard modular buoy body using polyethylene polymer material for ship safety. J Mater Sci Chem Eng 4(01):65–73
4. Piya-areetham P, Prasassarakich P, Rempel GL (2013) Organic solvent-free hydrogenation of natural rubber latex and synthetic polyisoprene emulsion catalyzed by water-soluble rhodium complexes. J Mol Catal A Chem 372:151–159
5. Idris MSF, Yusoff SFM, Mokhtar W (2019) New approach on the modification of liquid natural rubber production using microwave technique. Sains Malaysiana 48(7):1433–1438

6. Cifriadi A, Chalid M, Puspitasari S (2017) Characterization of hydrogenated natural rubber synthesized by diimide transfer hydrogenation. Int J Technol 8(3):448–457
7. Inoue S, Nishio T (2007) Synthesis and properties of hydrogenated natural rubber. J Appl Polym Sci 103(6):3957–3963
8. Puspitasari S, Falaah AF, Widiyantoro ANZ (2019) Selection of stabilizer and coagulant for natural rubber latex colloidal system during diimide catalytic hydrogenation at semi pilot scale reaction. IOP Conf Ser Mater Sci Eng:012128. IOP Publishing
9. Marsya MA, Putranto BD, Puspitasari S, Cifriadi A, Chalid M (2019) Catalyst screening on diimide transfer hydrogenation of natural rubber latex. IOP Conf Ser Mater Sci Eng:012078. IOP Publishing
10. Pereira AFP (2015) Colloidal stability of the latex particles. Instituto Superior Técnico
11. Mekonnen TH, Ah-Leung T, Hojabr S, Berry R (2019) Investigation of the co-coagulation of natural rubber latex and cellulose nanocrystals aqueous dispersion. Colloids Surf A 583:123949
12. Gregory J (2009) Monitoring particle aggregation processes. Adv Coll Interface Sci 147:109–123
13. Carlsson G (2004) Latex colloid dynamics in complex dispersions: fluorescence microscopy applied to coating color model systems. Karlstad University
14. Instruments Malvern (2016) A basic introduction to rheology. malvern instruments limited, Worcestershire, UK

Applied Chemistry and Chemical Engineering

Effect of Temperature on the Dehydration of the Pulp of Three Ecotypes of the Species *Mauritia Flexuosa* (*Aguaje*), on the Antioxidant Capacity, Total Polyphenols, and Anthocyanins of the Oil Extracted by Cold Pressure, Ucayali-Perú

C. Ruiz, D. Arancibia, S. Camargo, A. Da Cruz, E. Canahuire, G. Camargo, and W. Llatance

1 Introduction

In the Peruvian Amazon, there are countless species with great potential for agribusiness, which have been little, or nothing exploited, including the *Mauritia flexuosa* (Arecaceae), which is a palm tree that lives on the banks of rivers, streams, lakes, and springs. It has an elliptical or oval fruit, surrounded by a bark of triangular scales of reddish-brown color, with a thin, orange, fleshy, and oily mesocarp, which is used considerably in the manufacture of varieties of products [1].

Aguaje oil extraction is known for its functional properties due to the high concentrations of monounsaturated fatty acids; the amounts are higher than olive and

C. Ruiz (✉) · D. Arancibia
Universidad Nacional de Ucayali, Pucallpa, Perú
e-mail: carlos_ruiz@unu.edu.pe

S. Camargo
Universidad Continental, Huancayo, Perú

A. Da Cruz
Universidad Nacional Agraria de La Selva, Tingo Maria, Perú

E. Canahuire
Universidad Nacional Jorge Basadre Grohmann, Tacna, Perú

G. Camargo
Universidad Peruana de Ciencias Aplicadas, Lima, Perú

W. Llatance
Universidad Nacional de Jaén, Jaén, Perú

© The Author(s), under exclusive license to Springer Nature Singapore Pte Ltd. 2023

S.-M. Chen (ed.), *Proceedings of 10th International Conference on Chemical Science and Engineering*, Springer Proceedings in Materials 21,
https://doi.org/10.1007/978-981-19-4290-7_5

Brazilian walnut oils, recognized as oils high in bioactive compounds [2]. During the study, it was possible to determine the adequate temperature of dehydration of the pulp of the three ecotypes of aguaje (amarillo, ponguete, and shambo), obtaining dry pulp to obtain oil in cold extraction, where its bioactive compounds were analyzed to determine if there were some variations when working with two temperatures [3], with these results it is desired to provide transformation proposals and thus generate an agro-industrial alternative, that allows to give it an added value and increase the economy of the producer [4].

2 Materials and Methods

2.1 *Place of Execution*

The fulfillment of the research work was developed at the National University of Ucayali located with UTM coordinates (Mercator Transversal Universal) 18L 546,463 9,072,264, next to the Federico Basadre Highway Km 6200. The extraction of aguaje oil was carried out in the Center for Training and Elaboration of Medicinal and Food Plants of the Faculty of Agricultural Sciences where the Oil and Fat Technology Workshop is located, while the physicochemical analyses were carried out in the specialized laboratory of fruits and vegetables of the professional school of agro-industrial engineering.

2.2 *Raw Material*

The ecotypes of the species *Mauritia flexuosa* (amarillo, ponguete, and shambo) were collected in the provinces of Coronel Portillo and Padre Abad and then acquired in the following places.

2.2.1 Amarillo

This ecotype was acquired in the populated center "Santa Rosa" province of Coronel Portillo, District of Campo Verde km 47, entering the right bank km 4600. Of the Federico Basadre Highway, with coordinates of the palm tree UTM 18L 511,890 9,061,599.

2.2.2 Ponguete

This ecotype was acquired in the town center "El Pimental" of the province of Coronel Portillo, District of Campo Verde km 34, entering the left bank km 5, from the road to Tournavista, with coordinates of the palm tree UTM 18L 521,402 9,056,202.

2.2.3 Shambo

This ecotype was acquired in the town center "El Boquerón" km 147 of the CFB district of Aguaytia, Shambillo hamlet interior right bank 6 km. with the coordinates of the UTM 18L palm 432,988 9,003,923.

2.3 *Materials and Reagents*

2.3.1 Reagents

Distilled water, hexane, HCl (0.1 N, 1 N; 0.05 N), NaOH (1 N, 0.1 N), ethanol (60%), sodium hydroxide 0.1 N, phenolphthalein, hydrochloric acid 0.1 N.

2.3.2 Materials

Burette (25 ml), glass rod, test tubes, pipettes (10 ml), porcelain capsules, spatula, tweezers, magnets, vessels, precipitates (50, 100, 250, 600 ml), pizeta, graduated sample (25, 50, 100, 250 ml), fiola (250 ml), thermometer, Buchner funnels, Erlenmeyer flask (250 ml), test tube grids, test tube brushes, filter paper, parafilm, kraft paper, universal stand, plastic bucket, 240 ml amber bottles, laboratory accessories (apron, cap, mask, and gloves).

2.4 *Experimental Methodology*

2.4.1 Physicochemical Characterization

For the determination of the proximal chemical analysis, the AOAC method was used, this was performed at both temperatures ($T1 = 50$ °C and $T2 = 60$ °C) [5].

2.4.2 Determination of Antioxidant Capacity

The determination of the antioxidant capacity was carried out by the DPPH method (2,2-diphenyl-1-picrilhidrazilo) inhibition of the radical 2,2-difenil-1-picrilhidrazilo (DPPH°), described by Brand et al. [6].

2.4.3 Determination of Total Polyphenols

The determination of total polyphenols was performed by the spectrophotometric method developed by Folin Ciocalteu reported by with some modifications [7].

2.4.4 Determination of Anthocyanins

The quantification of anthocyanins was performed using the differential pH method reported by Rapisarda et al. [8].

2.5 *Statistical Design of the Research*

2.5.1 Experimental Design

The completely random design (DCA) was used, with 3 × 2 factorial arrangement with three repetitions. We then proceeded to use the nonparametric analysis: Analysis of variance and Tukey's multiple mean comparison test is found in Fig. 1.

a) **Factor and levels.**
   The focus of the study factor and its different levels is found in Table 1.

b) ***Experimental Scheme***

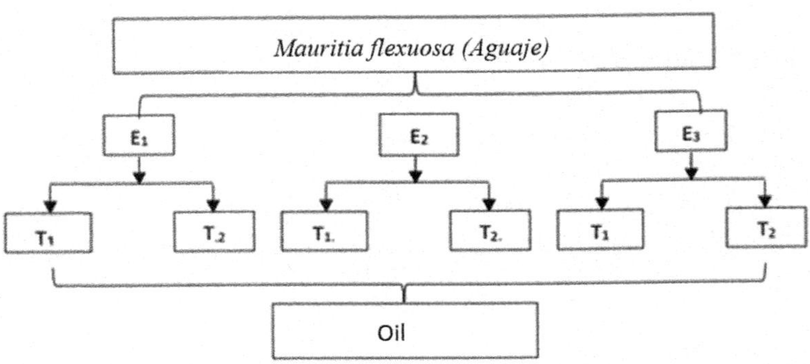

Fig. 1 Factor and level of experiment

Table 1 Study factor and its different levels

Factor	Temperature	Ecotypes
Aguaje oil	50 °C	Amarillo
		Ponguete
		Shambo
	60 °C	Amarillo
		Ponguete
		Shambo

E_1 Ecotype 1 (Shambo)
E_2 Ecotype 2 (Ponguete)
E_3 Ecotype 3 (Amarillo)
T_1 50 °C
T_2 60 °C

2.6 *Statistical Model*

2.6.1 Evaluation of Proximal Chemical Characteristics

The statistical model was:

$$Y_{ijk} = \mu + P_i + M_j + (P \times M)_{ij} + E_{ijk}$$

where

Y_{ijk} Observation (physicochemical characteristic)
M General population average
P_i General average of the population
M_j Effect of pulp conditioning (dehydration temperature 50 °C and 60 °C)
$(P \times M)_{ij}$ Effect of the interaction of the zones with the collection time
E_{ijk} Experimental error

3 Results

3.1 *Biometric Characteristics of the Three Ecotypes of Aguaje*

3.1.1 Longitudinal and Transverse Diameter of the Aguaje

The physical characteristics of 20 units of aguaje were evaluated for each of the three ecotypes taking as a result that the shambo ecotype is the aguaje with a superior length with a value of 6.44 cm, while for the diameter of the fruit is the upper aguaje ponguete with a measure of 3.84 cm, within its physical characteristics the aguaje shambo has a seed of greater size than the different ecotypes, with a measure of 4. 48 cm for seed length and 2. Sixty-five cm for the diameter of your aguaje seed ponguete are found in Fig. 2.

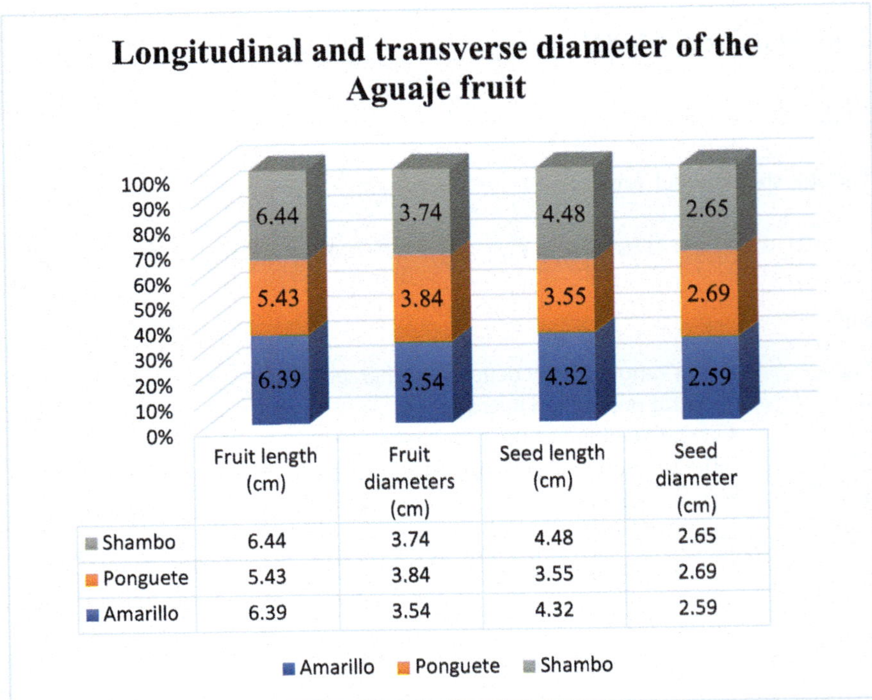

Fig. 2 Measurements of the fruits of the aguaje

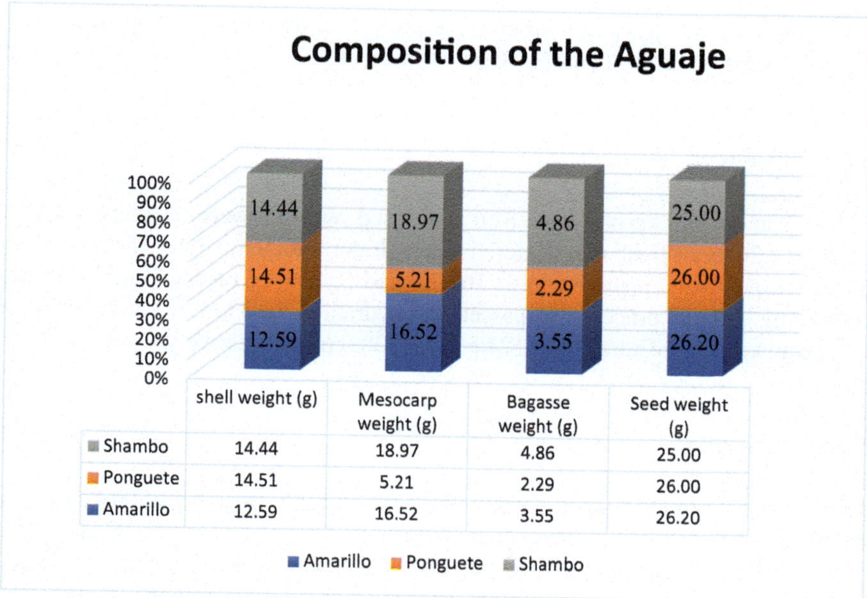

Fig. 3 Composition of aguaje

3.1.2 Composition of the Fruit of the Aguaje

The physical characteristics (weight) of 20 units of aguaje were evaluated for each of the three ecotypes, resulting in the highest weight of the shell being the aguaje ponguete with an amount of 14.51 g, while the shambo ecotype with the highest amount in the weight of the mesocarp with approximately 18.97 g, being the same one that has the highest weight of bagaza with an amount of 4.86 g, within the weight of the seed the marillo aguaje has 26.20 g. this information can be found in Fig. 3.

3.2 Oil Extraction

For the extraction of the oil from the pulp without aguaje shell with its three ecotypes at temperatures of 50 °C and 60 °C, a hydraulic press of 20 Tn was used giving the following results:

3.2.1 Shambo

42.25 kg and 39.10 kg of aguaje fruit were used where 9.20 kg and 7.40 kg of fresh mass were obtained that entered a dehydrator with temperatures of 50 °C and 60 °C, respectively, resulting in 4.40 kg and 3600 kg of dry mass of aguaje; continuing with the process was passed to a hydraulic press for the extraction of oil in cold aguaje that resulted in 419.10 ml and 542.0 ml, respectively.

3.2.2 Ponguete

37.85 kg and 43.40 kg of aguaje fruit were used, where 3 were obtained. Hundred kg and 12.20 kg of fresh mass entered a dehydrator with temperatures of 50 °C and 60 °C, respectively, resulting in 3. 100 kg and 5. 700 kg of dry mass of aguaje, continuing with the process was passed to a hydraulic press for the extraction of oil in cold aguaje that resulted in 1. 127 ml and 928.0 ml, respectively.

3.2.3 Amarillo

Thirty-five and 44 kg of aguaje fruit were used where 8700 and 11,500 kg of fresh mass were obtained that entered a dehydrator with temperatures of 50 °C and 60 °C, respectively, resulting in 2700 kg and 3100 kg of dry mass of aguaje; continuing with the process was passed to a hydraulic press for the extraction of aguaje oil that resulted in 568 ml and 312. 7 ml, respectively.

3.3 *Physicochemical Analysis of Marillo, Shambo and Ponguete Ecotypes*

Physicochemical analyses of the three ecotypes of aguaje: marillo, shambo, and ponguete with two temperatures 50 and 60 °C are found in Table 2.

Table 2 Physicochemical analysis of aguaje oil

Analysis	Amarillo		Ponguete		Shambo	
	50 °C	60 °C	50 °C	60 °C	50 °C	60 °C
Humidity (%)	0.09	0.09	0.10	0.10	0.09	0.10
Density (g/ml)	0.918	0.908	0.906	0.907	0.917	0.918
Acidity (%)	1.29	1.48	3.72	2.65	3.38	2.58
Peroxide (meq O2/Kg)	9.10	11.55	10.28	9.34	9.97	10.79
Saponification (%)	185.53	185.57	187.01	187.18	181.09	190.07

Table 3 Antioxidants of aguaje oil with its two temperatures

Echotype	Repetition	Temperature	
		50 °C	60 °C
Amarillo	R1	27775.50	36683.80
	R2	32229.65	32724.56
	R3	33714.37	32003.32
Ponguete	R1	69872.50	32497.90
	R2	51658.30	39498.83
	R3	45586.90	39735.60
Shambo	R1	48250.00	33444.10
	R2	40373.95	49634.50
	R3	37748.60	50626.96

3.4 *Antioxidant Content*

The results obtained for the variables regarding the antioxidant capacity in obtaining aguaje oil show that there are no statistical differences between the main effects of the ecotypes and the temperature or interaction between them are found in Table 3.

3.5 *Total Polyphenol Content*

No statistical differences were found in the effect of temperature due to having a P value greater than 0.05; however, between the effects of the ecotypes, we found significant statistical differences for having a P value less than 0.05 proceeding to the Tukey test.

4 Discussion

As Vásquez expresses in his research work, they mention that the weight of the mesocarp (pulp) of the amarillo, ponguete and shambo ecotype is between 16.36 g, 14.50 g, and 18.97 g, respectively, while the length of the fruit of the aguaje is 6.46 cm, 5.45 cm, and 6.45 cm, respectively. Making a comparison with our results, the weight of the mesocarp (pulp) of the ecotype amarillo, ponguete and shambo is 16.14 g, 14.51 g, and 19.02 g, respectively, and for the length, we have 6.39 g, 5.43 g, and 6.44 g, respectively. Being the shambo ecotype with the highest yield of the mesocarp and for the length, the three ecotypes are in the parameters reported by the author [9].

5 Conclusion

The dehydration temperatures of 50 and 60 °C of the pulp influence the total polyphenol content of the oil extracted by cold pressure from the amarillo, ponguete, and shambo ecotypes; however, for antioxidant capacity and anthocyanin content, no effects were found on dehydration temperature or ecotypes. Thus, the oil extracted by cold pressure of the ecotypes studied has antioxidant potential on average such as amarillo (32,521.8667 μmol trolox/100 g), ponguete (46,475.005 μmol trolox/100 g), and shambo (38,346.3517 μmol trolox/100 g). Therefore, the dehydration temperature of 50 and 60 °C of the pulp of the ecotypes had no effect on the physicochemical parameters (density, humidity, acidity, saponification, and peroxides) in the oil extracted by cold pressure.

References

1. Ferreira BS, Almeida CG, Faza LP, Almeida A, Diniz CG, Silva VL (2005) Comparative properties of Amazonian oils obtained different extraction methods. Molecules 16:5875–5885
2. Sampaio M, Carrazza L (2012) Manual tecnológico de aproveitamento integral do fruto e da folha do buriti (Mauritia flexuosa), Brasilia
3. Delgado C, Couturier G, Mejia K (2007) Mauritia flexuosa (Arecaceae: Calamoideae) an Amazonian palm with cultivation purposes in Peru 62:157–169
4. Viera FC, Pierre C, Castro HF (2005) Influência da composição em ácidos graxos de diferentes óleos vegetais nas propriedades catalíticas de uma de uma preparação comercial de lípase pancreática., Campinas: VI Congresso Brasileiro de Engenharia química em Iniciação Científica
5. AOAC. Official methods of analysis, Association of official analytical chemists (2008)
6. Brand WW, Cuvelier M, Berset C (1995) Use of a free radical method to evaluate antioxidant activity 16:25–30
7. Sultana B, Anwar F (2009) The antioxidant activity of R. phenolic components present in bark barks of Azadirachta indica, Arjuna Terminalia, nilotica Acacia, and Eugenia jambolana Lam. trees 104:1106–1114
8. Rapisarda P, Fanella F, Maccarone E (2000) Reliability of analytical methods for determining Anthocyanins in blood orange juices. J Agric Food Chem 48:2249–2252
9. Vasquez P, Sotero V, Castillo D, Alvarado L, Maco M (2009) Chemical differentiation of three morphotypes of Mauritia flexuosa L.f of the Peruvian Amazon

Effect of Blanching Time and Par-Frying Temperature on Quality of Frozen Par-Fried Taro

P. Penjumras, S. Kunkrathok, S. Umnat, P. Chokeprasert, R. Pokkaew, I. Wattananapakasem, L. Naloka, and A. Phaiphan

1 Introduction

Taro (*Colocasia esculenta* L.) is called in different names including cocoyam, dasheen, tannia, or eddoe [1]. It is a source of calories because it is rich in carbohydrates (4.2–4.4 kcal/g dry matter) [2]. In addition, these tubers provide nutritionally beneficial chemical components such as mucilage and resistant starch. Resistant starch has been qualified with a slow digestion in the lower parts of the human gastrointestinal tract and aids in the reduction of obesity risk and glucose absorption [1]. It has high contents of magnesium and potassium and (118–219 mg/100 g dry matter and 2251–4143 mg/100 g dry matter, respectively [1, 2]. Most of the previous studies on taro have focused on the characterization of thermal, physicochemical, and microstructural properties of flours, starches, and pastes. Taro is a root crop which lacks utilization in terms of economic value. Its usage could be increased by developing optimum processing technology and securing consumer preference. Therefore, the present study was taken up to develop the product from taro tubers. Frozen par-fried taro product was selected for this study. Generally, the quality of frozen vegetable products is mainly based on conditions of processing like blanching, frying, and freezing [3]. Blanching is an unit operation prior to freezing,

P. Penjumras (✉) · S. Kunkrathok · S. Umnat · P. Chokeprasert · R. Pokkaew · I. Wattananapakasem
Program of Food Technology, Maejo University-Phrae Campus, Phrae 54140, Thailand
e-mail: patpenp@gmail.com

L. Naloka
Program of Crop Production Technology, Maejo University-Phrae Campus, Phrae 54140, Thailand

A. Phaiphan
Program of Food Technology, Faculty of Agriculture, Ubon Ratchathani Rajabhat University, Ubon Ratchathani 34000, Thailand

© The Author(s), under exclusive license to Springer Nature Singapore Pte Ltd. 2023
S.-M. Chen (ed.), *Proceedings of 10th International Conference on Chemical Science and Engineering*, Springer Proceedings in Materials 21,
https://doi.org/10.1007/978-981-19-4290-7_6

drying, canning, and others processed food produced from fruit and vegetables for the aim of enzymes browning prevention to protect color and lowering amount of the reducing sugar in raw material [4], modifying texture, flavor and nutritional value, and trapped air removal [5]. Frying condition is also the main factor affecting the quality of product [6]. This unit operation consists of the immersion of the food materials in hot edible oils above the boiling point and then produces desired quality behavior such as color, flavor, texture, and mouthfeel [2]. The application of blanching and frying treatments for potato has been reported by Agblor and Scanlon [7] on the well-known French fry cultivars, Russert Burbank, and Shepody. That study demonstrated that blanching at 97 °C for 2 min and frying at 182 °C for 1.5 min produced fries with a lower crust hardness, whereas blanching at 70 °C for 10 min and frying at 166 °C for 3.5 min produced the contrast data. The effect of different partial-frying time and temperature combinations, on the cultivar, Agria, was observed [8] and presented the similar results. To date, frying of starch foods has mainly been studied for potatoes. Other tubers such taro have not been widely explored yet. Therefore, this study focuses on the effect of blanching time and partial-frying temperature on behavior of frozen par-fried taro to find the alternative process for taro utilization.

2 Materials and Methods

2.1 Raw Material Preparation

Taro (*Colocasia esculenta* L.) was purchased from the local market in Phrae Province (Thailand). Hundred percentage refined soybean oil (Angoon brand, Thai Vegetable Oil Public Company Limited) was used for frying in this study. Taro tubers were washed thoroughly with tap water, peeled by hand using a knife, and then cut into strips of 1 × 1 × 7 cm and immediately rinsed with tap water to remove surface starch and inhibit enzymatic browning reaction.

2.2 Study of the Effect of Blanching Time and Frying Temperature

A completely randomized design (CRD) was conducted to evaluate the effect of blanching time (0, 5, and 10 min) on the quality of frozen par-fried taro. The taro strips were blanched in hot water at 85 °C and then soaked in cooled water at 4 ± 2 °C for 1 min. The strips were then removed from the excess water. The strips were par-fried in hot refined soybean oil at 180 °C for 2 min using the taro-to-oil ratio at 1 g: 40 ml. After frying, the strips were allowed to drain and cool for 5 min at room temperature (25 °C). Each treatment was packed and kept at –20 ± 2 °C for 7 days before testing. The selected blanching time was used for further study. The effect of frying temperature (160, 180, and 200 °C) for 2 min was employed by a completely randomized design (CRD).

2.3 *Method of Analytical*

The preparation of the sample for analysis is as follows: thawing at room temperature for 15 min then frying in hot soybean oil of 180 °C for 3 min. The samples were then analyzed.

Color Measurement. Color of taro strips was measured within 30 min of frying using Hunter Lab (Color Flex 500, Hunter Lab, USA). Results were expressed in CIE color values; L* = lightness (0 = black, 100 = white), a* (−a* = greenness, +a* = redness), and b* (−b* = blueness, +b* = yellowness).

Moisture Content and Lipid Content. Moisture content and lipid content were determined according to AOAC (2012) [9]. All the analyses were performed in triplicates.

Sensory Evaluation. Sensory attributes were evaluated by 30 panelists who are the members of Food Technology Program, Maejo University-Phrae Campus, Thailand. The level of preference for taro strips was rated by a 9-points hedonic scale test. All untrained panels were informed to evaluate samples. The 30 panelists received samples and were asked to rate them based on level of preference on a 9-points hedonic scale (1 = dislike extremely, 4 = dislike slightly, and 9 = like extremely) to evaluate the product characteristics (appearance, color, flavor, taste, texture, and overall acceptance) [10].

2.4 *Statistical Analysis*

Data were conducted to Analysis of Variance (ANOVA) using SPSS for Window version 24. In case of any differences in mean, multiple comparisons were performed using Duncan's Multiple Range Test (DMRT) at 5 % level of significance ($P \leq 0.05$).

3 Results and Discussion

3.1 *The Effect of Blanching Time on Quality of Taro Strips*

The effect of blanching time at 85 °C on physical and chemical properties and sensory characteristics of taro strips were investigated. The results are presented in Tables 1 and 2.

Table 1 Effect of blanching time on color, moisture content, and lipid content of taro strips

Properties	Blanching time (min)		
	0	5	10
Color Lightness (L*)	27.09 ± 0.90[a]	37.51 ± 0.35[b]	36.82 ± 0.50[b]
Redness (a*)[ns]	−2.07 ± 0.36	−1.40 ± 0.21	−1.42 ± 0.69
Yellowness (b*)[ns]	8.43 ± 0.30	8.45 ± 0.34	8.46 ± 0.28
Moisture content (%)	41.87 ± 0.08[c]	31.36 ± 1.36[b]	27.67 ± 1.56[a]
Lipid content (%)	11.18 ± 0.82[a]	13.51 ± 0.53[b]	13.08 ± 0.29[b]

Mean ± standard deviation values followed by a different letter within the same row are significantly different ($p \leq 0.05$) by Duncan's multiple range test.
ns not significant ($p > 0.05$) different within the same row by Duncan's multiple range test.

Table 2 Effect of blanching time on sensory characteristics of taro strips

Characteristics	Blanching time (min)		
	0	5	10
Color[ns]	7.00 ± 1.46	7.07 ± 1.53	7.10 ± 1.73
Flavor[ns]	7.26 ± 1.36	7.13 ± 1.20	7.33 ± 1.45
Taste[ns]	6.87 ± 1.31	7.33 ± 1.75	7.43 ± 1.68
Texture	6.03 ± 1.77[a]	7.23 ± 1.72[b]	7.10 ± 2.11[b]
Overall acceptance[ns]	6.80 ± 1.58	7.40 ± 1.73	7.47 ± 1.76

Mean ± standard deviation values followed by a different letter within the same row are significantly different ($p \leq 0.05$) by Duncan's multiple range test.
ns not significant ($p > 0.05$) different within the same row by Duncan's multiple range test.

From Table 1, ANOVA shows significantly different physical properties of lightness (L*). The lightness of strips was found to increase with increasing blanching time. This could be related to inactivation of polyphenol oxidase and peroxidase which are the cause of enzymatic browning [4] but the redness (a*) and yellowness (b*) values of taro strips were not significantly different, affected by blanching time. This result is in agreement with previous research that blanching in hot water would inhibit browning reaction as the result of polyphenoloxidase activity [11]. However, longer time in blanching shows negative impact on texture. This is probably due to overheating accumulation in starch and texture becoming softer than the strip broken.

This study, two chemical properties consisting of moisture content and lipid content, was measured. Moisture evaporation from food products during immersion frying damages the cellular structure of plant tissues causing increased porosity which contributes to oil absorption [12]. The results in Table 1 show that moisture content and lipid content were significantly ($p \leq 0.05$) different among treatment resulting from blanching time. The increase of blanching time affects decrease of moisture content but increase in lipid content. However, there was no difference in lipid content between treatment of 5- and 10-min blanching. The results found in this recent study contrast with previous study. [4] who reported that French fry moisture content did not significantly ($p \leq 0.05$) differ by blanching treatment. In addition, [13] found that moisture content of blanched potato flour was significantly superior compared with unblanched potato flour, probably due to water absorbed by potato during blanching. However, in this present study, the tubers used were taro that generally contains hydrocolloid like gum and mucilage with considerable potential as a food thickening agent [1]. Thus, this thickener could block removal of water from material and offer higher moisture content compared to blanched samples. The results were similar with [12] who found that moisture content of coated strips with hydrocolloid such as CMC, pectin, and agar before processing was significantly higher than that of uncoated strips. Sensory characteristics were tested to evaluate acceptance by the untrained panel as shown in Table 2.

Table 2 demonstrates that there was a significant ($p \leq 0.05$) difference in texture score affected by blanching compared with unblanched taro strips. The blanched samples show superior score in texture. Although there were no significant differences in color, flavor, taste, and overall acceptance, the score tends to increase due to blanching treatment. [5] informed that blanching could modify texture flavor and nutritional value, and removing trapped air then provides a positive impact to strips. Therefore, the blanching treatment at 85 °C for 5 min should be the selected condition for further steps. The effect of par-frying temperature on properties of taro strips was evaluated. The results were represented in Tables 3 and 4.

Table 3 Effect of par-frying temperature on color, moisture content, and lipid content of taro strips

Properties	Par-frying temperature (°C)		
	160	180	200
Color Lightness (L^*)^{ns}	42.65 ± 1.55	44.40 ± 0.87	44.99 ± 2.38
Redness (a^*)	−0.48 ± 0.52^{a}	−0.48 ± 0.52^{a}	−1.72 ± 1.30^{ab}
Yellowness (b^*)	10.79 ± 0.04^{a}	10.71 ± 0.25^{b}	10.06 ± 0.48^{ab}
Moisture content (%)^{ns}	46.85 ± 1.03	46.01 ± 0.82	46.75 ± 1.07
Lipid (%)	20.72 ± 0.67^{a}	7.19 ± 2.14^{b}	7.25 ± 1.08^{b}

Mean ± standard deviation values followed by a different letter within the same row are significantly different ($p \leq 0.05$) by Duncan's multiple range test.
ns not significant ($p > 0.05$) different within the same row by Duncan's multiple range test.

Table 4 Effect of par-frying temperature on sensory characteristics of taro strips

Characteristics	Par-frying temperature (°C)		
	160	180	200
Color	5.95 ± 1.60[a]	6.50 ± 1.84[ab]	7.20 ± 1.00[b]
Flavor	6.25 ± 1.58	6.55 ± 1.00	6.85 ± 1.28
Taste[ns]	6.60 ± 1.57	7.25 ± 1.02	7.45 ± 1.05
Texture	5.55 ± 1.88[a]	6.20 ± 2.26[ab]	6.95 ± 1.57[b]
Overall acceptance	6.10 ± 1.48[a]	6.80 ± 1.28[ab]	7.45 ± 1.02[b]

Mean ± standard deviation values followed by a different letter within the same row are significantly different ($p \leq 0.05$) by Duncan's multiple range test.
ns not significant ($p > 0.05$) different within the same row by Duncan's multiple range test.

3.2 *The Effect of Par-Frying Temperature on Quality of Strips*

Results in Table 3 reveal that par-frying temperature had no significant ($p \leq 0.05$) difference on lightness (L* value) and moisture content but provided an impact on redness (a*), yellowness (b*) values, and lipid content. The redness (a*) value of taro strips decreased with an increased par-frying temperature which indicated a decrease in yellowness. In addition, the results presented lower in oil content at higher frying temperature. This may be explained by the enhanced crust formation which acts as a physical barrier for oil absorption [2]. The results agree with previous study of [2]. Sensory characteristics were then evaluated as data shown in Table 4. An increase in par-frying temperature tends to increase sensory evaluation score. This is probably because at 160 °C par-frying temperature provided the highest lipid content and panelists disliked the oiliness of taro strips. In addition, at lower temperature of par-frying, the strips' texture may be sticky due to the swelling pressure of the hydrated starch granules might not have been enough to rupture cell walls [6]. The results presented a similar finding with the previous study by [4, 6]. However, the difference between 160 °C and 180 °C, and 180 °C and 200 °C were not significant ($p > 0.05$). Therefore, the suitable par-frying temperature for frozen taro could be 180 °C.

4 Conclusions

The results from this study indicate that blanching time had significant effects on lightness (L*), moisture, and lipid content, and taste and texture, meanwhile par-frying temperature had significant influence on redness (a*) and yellowness (b*) values, lipid content, and sensory characteristics except taste. In light of data presented in this study, it can be concluded that the most proper condition to produce frozen par-fried taro of the best quality was blanching at 85 °C for 5 min, soaked in cooled water and then removing the excess water. The strips were then par-fried in hot soybean oil at 180 °C for 2 min. After frying, then cooling and packing, the packed strips were then kept in the frozen system at -20 ± 2 °C. The present findings can be used for the development of frozen par-fried taro to be an alternative utilization of taro tubers.

Acknowledgements The authors would like to acknowledge Program of Food Technology, Maejo University, Phrae Campus, and for providing facilities all technical staff for their support.

References

1. Kaushal P, Kumar V, Sharma HK (2015) J Food Sci Technol 52:27
2. Paz-Gamboaa E, Ramírez-Figueroaa E, Vivar-Veraa MA, Bravo-Delgadoa HR, Cortés-Zavaletab O, Ruiz-Espinosac H, Ruiz-López II (2015) CyTA J Food 13:506
3. Elmore JS, Briddon A, Dodson AT, Muttucumaru N, Halford NG, Mottram DS (2015) Food Chem 182:1
4. Ngobese NZ, Workneh TS, Siwela M (2017) J Food Sci Technol 54:507
5. Kapadiya DC, Makavana JM, Kathiria MK (2018) Int J Curr Microbiol App Sci 7:2754
6. Millin TM, Medina-Meza IG, Walters BC, Huber KC, Rasco BA, Ganjyal GM (2016) Food Bioprocess Technol 9:2080
7. Agblor A, Scanlon MG (2000) Potato Res 43:163
8. van Loon WAM, Visser JE, Linssen JPH, Somsen DJ, Klok HJ, Voragen AGJ (2007) Eur Food Res Technol 225:929
9. AOAC (2012) Official method of analysis of AOAC International, vol 19. Association of official and analytical chemists international, Virginia
10. Penjumras P, Thongfathamrong P, Umnat S, Chokeprasert P, Wattananapakasem I, Phaiphan A (2021) IOP Conf Ser Earth Environ Sci 756(1)
11. Chen X, Lu J, Li X, Wang Y, Miao J, Mao X, Zhao C, Gao W (2017) LWT Food Sci Technol 82:303
12. Kizito KF, Abdel-Aal MH, Ragab MH, Youssef MM (2018) J Agric Sci Bot 1:18
13. do Nascimento RF, Canteri MHG (2018) Hortic. bras., Brasília 36:461

Adsorption Cationic Dye on Modified Chitosan with Sodium Dodecyl Sulfate

Chonlada Chumsing and Kowit Piyamongkala

1 Introduction

Textile and garment industry in Thailand is expanding rapidly with export value. It reached 2856.8 million US dollars in the first half of 2020 [1]. There are many fabric bleaching factories that using the water as solvent dissolved dyes which dye cannot be digested naturally. Both the intensive color concentration and high turbidity are main problem of wastewater from the dyeing industry. Synthetic dyes are hydrocarbon, aliphatic and aromatic chemicals. It is very difficult to remove or decompose from itself structure. The dye particles are very small; thus, it is complex to precipitate the pigment in the wastewater. Releasing of wastewater from the dyeing factory into natural water sources is event that cannot happen. It affects the livelihood of living things in water resources. Therefore, wastewater must be treated before discharge to water sources [2].

The treatment of wastewater resulting from fabric bleaching processes can be carried out in many ways, including activated carbon [3], precipitation [4], chemical oxidation [5], ozonation [6], membrane filtration [7], biodegradation [8] and electrochemical techniques [9]. The above methods have limited in wastewater treatment, due to the dye has inert properties to the action and low concentration. This results in high costs for wastewater treatment [10]. Several methods have been studied recently for the development of wastewater treatment with cheaper and more efficient adsorption such as chitosan [11]. It contains the large number of active hydroxyl groups (–OH) and amines ($-NH_2$); thus, it can absorb anionic dyes well [12]. However,

C. Chumsing (✉) · K. Piyamongkala
Department of Industrial Chemistry, Faculty of Applied Science, King Mongkut's University of Technology North Bangkok, Bangkok, Thailand
e-mail: Chonlada.ccc@hotmail.com

K. Piyamongkala
e-mail: kowit.p@sci.kmutnb.ac.th

© The Author(s), under exclusive license to Springer Nature Singapore Pte Ltd. 2023
S.-M. Chen (ed.), *Proceedings of 10th International Conference on Chemical Science and Engineering*, Springer Proceedings in Materials 21,
https://doi.org/10.1007/978-981-19-4290-7_7

using chitosan as the adsorbent requires development to optimize for each type of dye adsorption. Modifying the chitosan shows effective in absorbing the positively charged dyes and increasing the acid-resistance property. Sodium dodecyl sulfate (SDS) is an anionic surfactant. It was used in the modification of chitosan for the adsorption of cationic dyes [13]. Sodium dodecyl sulfate decreases the amine group density in the chitosan and affects the electric charge, which neutralizes the cations of the chitosan molecule [14]. Chitosan modified with sodium dodecyl sulfate has an increased anion. It is the result of the hydrophobic interaction between the dye molecules and the surfactant [15]. The dyeing industry, silk, cotton, leather and fur uses a basic dye such as methylene blue [16]. Methylene blue is moderately toxic to animals; 50 percent of rats were killed when ingested 1,180 mg of methylene blue per 1 kg of body weight [17]. Although toxicity is not much, ingestion in human body can cause unwanted symptoms such as rapid heartbeat, vomiting, shock, green skin due to lack of oxygen (cyanosis), jaundice, quadriplegia and necrosis [18]. Long-term exposure to methylene blue can lead to cancer or genetic mutations in humans and other organisms. How to effectively eliminate or replace methylene blue has become important to society and the environment [19].

This study investigated adsorption of methylene blue using the modified chitosan with sodium dodecyl sulfate. The modified chitosan was studied with sodium dodecyl sulfate concentrations. Adsorption efficiency and resistance of acid and base were studied in suitable solution.

2 Experimental

2.1 Chemical

Powder chitosan (degree of deacetylation of 93%) was purchased from Sin Udom Co., Ltd., Thailand. The laboratory-grade chemicals included sodium dodecyl sulfate and methylene blue, while the analytical grade consisted sodium hydroxide, acetic acid, hydrochloric acid and sulfuric acid.

2.2 Forming Modified Chitosan Beads with Sodium Dodecyl Sulfate

About 2 g of powder chitosan (PW) was added in the 2 v/v% of acetic acid solution. The mixture was stirred with stirrer at 300 rpm for 8 h to obtain the chitosan solution. Then, the chitosan solution was pumped by peristaltic pump at the flow rate 1.67 cm^3/min. It moved through the small tube and dropped into the 2.5, 5.0, 10.0, 15.0, 20.0 and 25.0 g/L of SDS to form chitosan beads. After that, it was filtered and washed with screening and distilled water to obtain modified chitosan beads. They were placed on the plastic tray to achieve the dry modified chitosan beads (MCSDS).While forming chitosan beads with sodium hydroxide showed the method as the dry modified chitosan beads but, the sodium dodecyl sulfate solution was replaced with 1 M of the sodium hydroxide to attained dry chitosan beads (CB).

2.3 Batch Adsorption Studies

The batch adsorption of dry modified chitosan beads, dry chitosan beads and powder chitosan was used to study the equilibrium adsorption experiments. Then, 0.2 g of each adsorbent was added in the Erlenmeyer flask and then 100 cm^3 of 100 mg/L of methylene blue solution was supplied into the Erlenmeyer flask too. The mixture was shaken by shaker at 120 rpm for 5 h in the room temperature. After that, the sample was withdrawn from Erlenmeyer flask. The initial and residual concentrations of methylene blue solution were analyzed by spectrophotometer at wavelength 665 nm. The adsorption capacity and percentage dye removal were calculated according Eqs. (1) and (2), respectively. The pH of solution was measured both before and after adsorptions by pH meter.

$$q_e = \frac{\left(C_0 - C_{\mathrm{eq}}\right) \times V}{W} \quad (1)$$

$$\text{Percentage dye removal}(\%) = \frac{\left(C_0 - C_{\mathrm{eq}}\right)}{C_0} \times 100 \quad (2)$$

where q_e is the adsorbent capacity (mg/g), C_0 is the initial concentration of methylene blue (mg/L), C_{eq} is the equilibrium concentration of methylene blue (mg/L), V is the volume of solution (L), and W is the weight of adsorbent (g).

2.4 Weight Loss

Weight loss of particles was characterized by determining the mass loss due to dissolution in distilled water, acid and base solutions. About 0.2 g of each adsorbents was supplied in the Erlenmeyer flask, then 100 cm^3 of basically included 0.1 M sodium hydroxide, acidly consisted 0.1 M hydrochloric, 0.1 M sulfuric acids, 0.1 M acetic acid and distilled water were presented in the Erlenmeyer flask too. The mixture was shaken by shaker at 120 rpm for 24 h in the room temperature. Adsorbent beads were filtered with filter paper and dry at 50 °C in the oven for 1 h. The percentage dry weight loss of the adsorbent beads was determined by the difference of dry weight of the adsorbent beads both before and after the drying.

3 Results and Discussions

3.1 Methylene Blue Adsorption Efficiency

High value of the adsorption capacity and the percentage dye removal indicate the good adsorption efficiency of adsorbent. Powder chitosan, dry chitosan beads (CB) and dry modified chitosan beads 2.5 g SDS (MCSDS) in the adsorption of methylene blue dye are shown in Fig. 1. The adsorption capacity was 7.8, 11.9 and 28.1 mg/g, respectively, while the percentage dye removal was 18.1, 20.5 and 57.3%, respectively. The modified chitosan 2.5 g SDS displayed the good efficiency due to the amino groups in the chitosan to form an aliphatic group resulting in more hydrophobic properties than chitosan [20]. The ionic compound effects of chitosan cause an intermolecular electrostatic attraction between an anionic surfactant and a cationic dye [21]. The adsorption capacity and percentage dye removal of methylene blue dye solution of modified chitosan with 2.5, 5.0 10.0, 15.0, 20.0 and 25.0 g of SDS are presented in Fig. 2. It was found that modified chitosan with 5.0 g of SDS

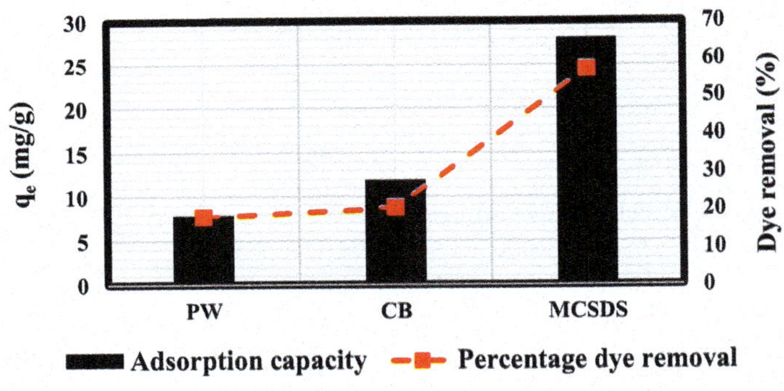

Fig.1 Adsorption of methylene blue by PW, CB and MCSDS

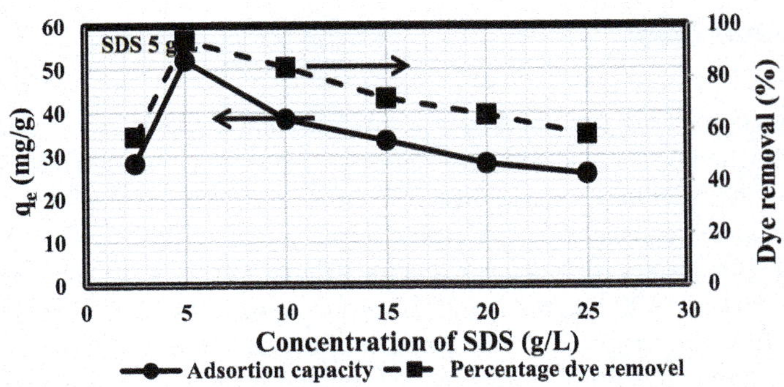

Fig. 2 Effect of sodium dodecyl sulfate concentration variation on adsorption capacity and percentage dye removal of methylene blue

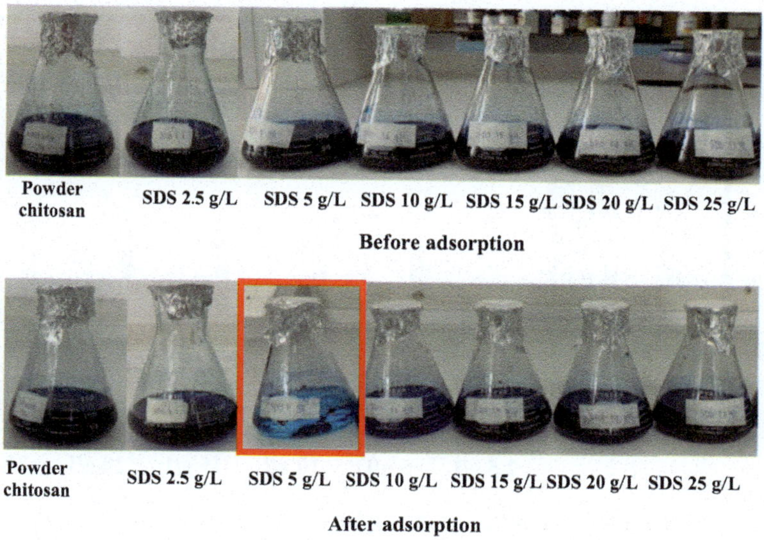

Fig. 3 Color of the methylene blue solution before and after absorption

showed adsorption efficiency higher than modified chitosan with 2.5 g of SDS due to surfactants which begin to bind to micelles. Most surfactants with hydrocarbon chains forming a non-polar structure in the molecule had a critical concentration of micelles in the range 10^{-4}–10^{-2} molar [22]. Surfactant molecules are bound to form micelles when the surfactant concentration is equal or higher than, then transformed into a free molecule before being converted back into a micelle form. Each micelle has a half-life 10^{-3} s [23]. The surface of the micelles is charged the same as the charges of the surfactants that bind to the micelles. The modified chitosan with SDS 5 g/L was in the range which the surfactant concentration was equal to the critical micelle concentration formation. Increasing amount of SDS higher than 5.0 g exposed decreasing in the adsorption efficiency due to the density of chitosan-modified SDS was increased. The molar ratio of SDS: amine groups in chitosan-modified SDS was only slightly increased with increasing SDS concentration; thus, the positive effect due to an increasing negative charge and hydrophobic interactions would not increase with increasing SDS concentration [13]. The color of the methylene blue dye before and after absorption is showed in Fig. 3. It is clearly seen that modified chitosan with SDS 5 g presented the brightest blue color of all adsorbents.

3.2 Weight Loss in Acid and Base Solutions

The adsorbent displays the low weight loss. It should be selected for using in the future due to the adsorbent which showed high resistance in variance of solution very well. The sodium hydroxide, hydrochloric acid, sulfuric acid, acetic acid and distilled

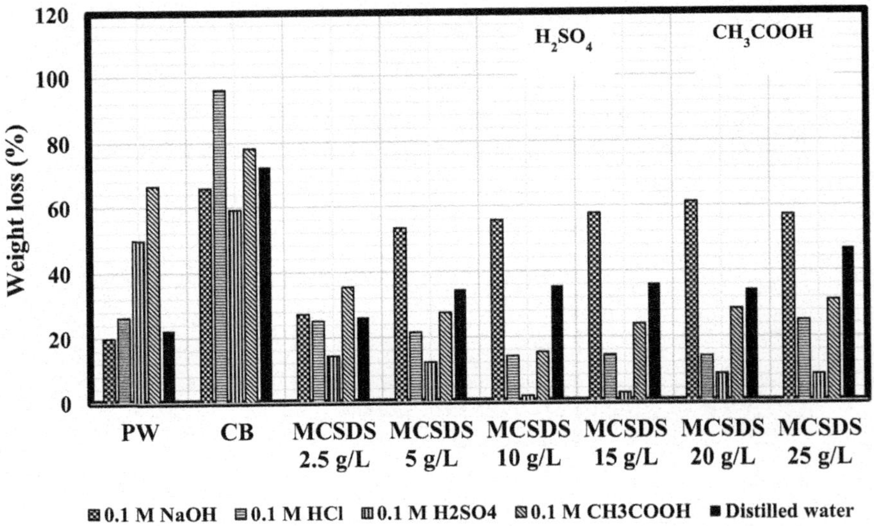

Fig. 4 Weight loss of adsorbent in acid and base solution

water are selected to test resistance of the adsorbents in solution and are shown in Fig. 4. Powder chitosan and dry chitosan beads cannot be kept at the shape of powder and bead in distilled water, sodium hydroxide and all acids condition. It presents the weight loss higher than 20%. The mass loss of powder chitosan and dry chitosan beads was found to increase with decreasing in pH of the acid solution. Because free positive charges develop in beads, the protonation of the amine groups of chitosan and the resulting mutual repulsion cause swelling. Dry modified chitosan bead formed by SDS did not show any mass loss or pH-sensitive swelling under acidic conditions [15]. Likewise, the shape of bead for modified chitosans was lost in distilled water and sodium hydroxide solution, while it failed a little bead in acetic acid solution. On the contrary, dry modified chitosan beads can be fixed in the appearance very well in hydrochloric acid and sulfuric acid solutions. It displayed the weight loss lower than 20%. The amino group of chitosan was crosslinked with sodium dodecyl sulfate, resulting modified chitosan stabilized in acid condition better. It means that dry modified chitosan beads can be used for the fine adsorbent for dyeing factory in which wastewater was contaminated with inorganic acid condition.

4 Conclusions

Methylene blue is a cationic dye, and it can be adsorbed by chitosan as adsorbents. The optimum sodium dodecyl sulfate concentration for the adsorption of methylene blue was 5 g/L. The concentration of sodium dodecyl sulfate is increasing, while the adsorption efficiency tended to be decreasing. The modified chitosan was stable in acid conditions than powder chitosan and dry chitosan beads.

References

1. Office of Industrial Economics Ministry of Industry (2020) Fashion Intelligence Unit
2. Tsai WT, Chang CY, Lin MC, Chien SF, Sun HF, Hsieh MF (2001) Chemosphere 45:51–58
3. Kadirvelu K, Kavipriya M, Karthika C, Radhika M, Vennilamani N, Pattabhi S (2003) Biores Technol 87:129–132
4. Hasani ZM, Alavi MMR, Arami M (2009) Water Sci Technol 59:1343–1351
5. Malik PK, Saha SK (2003) Sep Purif Technol 31:241–250
6. Selcuk H (2005) Dyes Pigm 64:222–271
7. Ahmad AL, Harris WA, S S, Ooi BS (2002) Journal Teknologi 36(1):31–44
8. Gopinath KP, Murugesan S, Abraham J, Muthukumar K (2009) Biores Technol 100:6295–6300
9. Radha KV, Sridevi V, Kalaivani K (2009) Biores Technol 100:987–990
10. Crini G, Badot PM (2008) Prog Polym Sci 33:399–447
11. Crini G (2006) Biores Technol 97:1061–1108
12. Elwakeel KZ (2010) J Dispersion Sci Technol 31:273–288
13. Chatterjee S, Chatterjee T, Woo SH (2010) Biores Technol 101(11):3853–3858
14. Onesippe C, Lagerge S (2008) Colloids Surf A 317:100–108
15. Chatterjee S, Lee DS, Lee MW, Woo SH (2009) Biores Technol 100:3862–3868
16. Kushwaha AK, Gupta N, Chattopadhyaya MC (2014) J Saudi Chem Soc 18(3):200–207
17. Seif C, Portillo FJM, Osmonov DK, Böhler G, Horst C, Leissner J et al (2004) Urology 63(6):1205–1208
18. Yi JZ, Zhang LM (2008) Biores Technol 99:2182–2186
19. Zhao G, Li C, Wu X, Yu J, Jiang X, Hu W et al (2018) Appl Surf Sci 434:251–259
20. Piyamongkala K, Mekasut L, Pongstabodee S (2008) Macromol Res 16(6):492–502
21. Fana S, Wanga Z, Wanga J, Tanga J, Li X (2017) Environ Chem Eng 5:601–611
22. Rosen MJ (1989) Wiley, p 431
23. Mittal KL, Fendler EJ (1980) Plenum Press

Enhancing Power Supply of Al-Air Battery Using an Optimized Conductive Material of Silica Xerogel/Graphite Composite on an Air Cathode

H. Aripin, Sutisna Sutisna, Nundang Busaeri, and Svillen Sabchevski

1 Introduction

The metal-air batteries are compact electrochemical power sources that have an anode of pure metal (e.g., Li, Na, K, Zn, Al, etc.), a cathode that uses oxygen from the ambient air, and a suitable (usually aqueous) electrolyte. They are considered as promising candidates for powering of various portable devices, electric vehicles, biosensors, etc., because their performance parameters such as power and energy density are higher than that of the lithium-ion batteries. Compared with other types of air metal batteries, the Al-air batteries are more attractive for various applications since they are characterized by a high-power density of about 8.1 kwh/kg, a high theoretical specific capacity of 2.98 Ah/g, and a high voltage of about 2.7 V [1]. In addition, the usage of Al anodes is economical and environmentally friendly because Al is cheap, abundant and can be recycled easily. During the redox reaction, however, the surface of the electrode is covered by the precipitate of $Al(OH)_3$ reaction products that impede the reaction that occurs between oxygen and water and the migration of OH^- to the Al anode. To suppress the accumulation of the reaction product on the cathode surface, it is necessary to coat it with an appropriate active material. The pore channels in such material are required to facilitate the repetition reactions that have a positive effect on increasing the discharge capacity, the efficiency of charging and discharging, and the life cycle of the battery. The performance of lithium-air batteries has been studied using carbon nanoparticles [2] and graphitic carbon nanofiber [3]

H. Aripin (✉) · S. Sutisna · N. Busaeri
Department of Electrical Engineering, Faculty of Engineering, Siliwangi University, Tasikmalaya, Indonesia
e-mail: aripin@unsil.ac.id

S. Sabchevski
Lab. Plasma Physics and Engineering, Institute of Electronics, Bulgarian Academy of Sciences Sofia, Sofia, Bulgaria

© The Author(s), under exclusive license to Springer Nature Singapore Pte Ltd. 2023
S.-M. Chen (ed.), *Proceedings of 10th International Conference on Chemical Science and Engineering*, Springer Proceedings in Materials 21,
https://doi.org/10.1007/978-981-19-4290-7_8

as the active material for air cathode. These investigations have demonstrated that the lithium-air batteries are characterized by high values of the specific energy above 1300 Wh/kg. However, the usage of lithium metal anodes poses well-known and serious safety problems. Attempting to overcome this hurdle, various magnesium-air batteries with different active materials have been proposed. Among the latter are MnO_2 [4], carbon [5], and MnO_2/C [6]. It has been found that such type of batteries is characterized by an even higher energy density of about 3100 Wh/kg. Regrettably, the magnesium-air batteries that use liquid electrolytes are affected by the corrosion of the Mg electrode, which results in electrochemical instability, especially during charging. As already mentioned above, another attractive type is represented by various aluminum-air batteries with different active materials, e.g., carbon nanotubes [7] and TiO_2 [8]. The cell voltage of these batteries is about 0.7 V with a maximum capacity of about 21.2 mAh/cm^2. Such values, however, are lower than that of magnesium- [5] and zinc [9]-air batteries. Consideration of these facts motivated us to look for an alternative air cathode material that could provide higher capacity.

Silica xerogel derived from various natural biomasses is an alternative active material whose pores are formed when the liquid phase is removed by evaporation. It has micro- to nano-size pores sizes, uniform and orderly pore arrangement and contains homogeneous siloxane groups. This material has been widely used for various engineering applications such as fuel cells [10], Zn-air battery [11], and lithium-air battery [12]. Silica, however, has low electrical conductivity of about 10^{-4} Sm^{-1} [13], so it is necessary to add conductive additive of graphite to improve the contact between the silica particles as well as between the current collector and the silica particles. This paper presents and discusses the results of recent experiments carried out to investigate the effects of the presence of SX in SX/graphite composite layer of an air cathode on the performance of an Al-air battery.

2 Methodology

2.1 Preparing the Layer of SX/Graphite

The amorphous SX, graphite, and polyvinylidene fluoride (PVDF) are used as material composite for an active layer in the air cathode. SX was prepared by an extracting method from sago waste ash. The extraction procedure has been described in detail elsewhere [14]. Using this technique, high-purity (about 98%) amorphous silica xerogel with an average particle size of 100–180 nm has been produced. Graphite and PVDF powder are commercially available materials, which were supplied from Sigma Aldrich. A composition on a dry basis of 2 wt.% SX and 93 wt.% graphite was mixed and then pulverized for 24 h at room temperature by an alumina ball mill. A ratio of 1:1 for PVDF and N-methylpyrrolidone (NMP) solution was used as a binder. Then, the milled SX/graphite was mixed with the binder and then stirred at

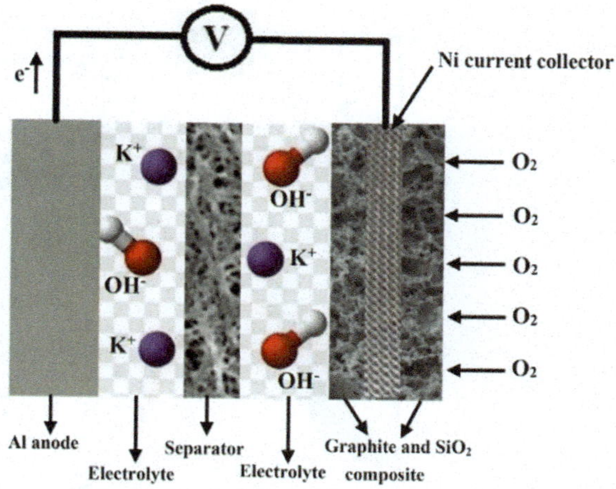

Fig. 1 Schematic drawing of the Al-air battery structure

room temperature for 30 min until the mixture became homogeneous slurry. After that, the slurry was ultrasonicated for at least 10 min to remove the air bubbles. The obtained slurry was coated on the substrate surface of a nickel 300 mesh and then was dried at room temperature. The mixtures containing SX of 8 wt.% (8 wt.% SX), 15 wt.% (15 wt.% SX), and 20 wt.% (20 wt.% SX) were prepared in a similar procedure as that described above.

2.2 *Method for Assembling and Characterizing the Al-Air Battery*

Figure 1 shows a schematic drawing of the Al-air battery structure. The rectangular battery cell is assembled with an anode and a cathode having sizes of 5×6 cm^2. An aluminum sheet and SX/graphite composite layer deposited on the surface of a nickel mesh substrate were selected as an anode material and an air cathode, respectively. The anode and cathode are separated by a porous separator membrane of tissue TRUWIPES. A 1 M KOH is used as an electrolyte, which was injected into the separator. The discharging capacity was tested using a battery testing system (BTS-MPTS, China) at a constant current density. The impedance data were tested by EIS at frequencies from 10^5 to 10^{-2} Hz and analyzed by using Nyquist plots.

3 Result and Discussion

Figure 2 shows Nyquist plot of EIS and the equivalent circuit model of the fitted data for Al-air batteries using an active layer at different SX contents in 1 M KOH

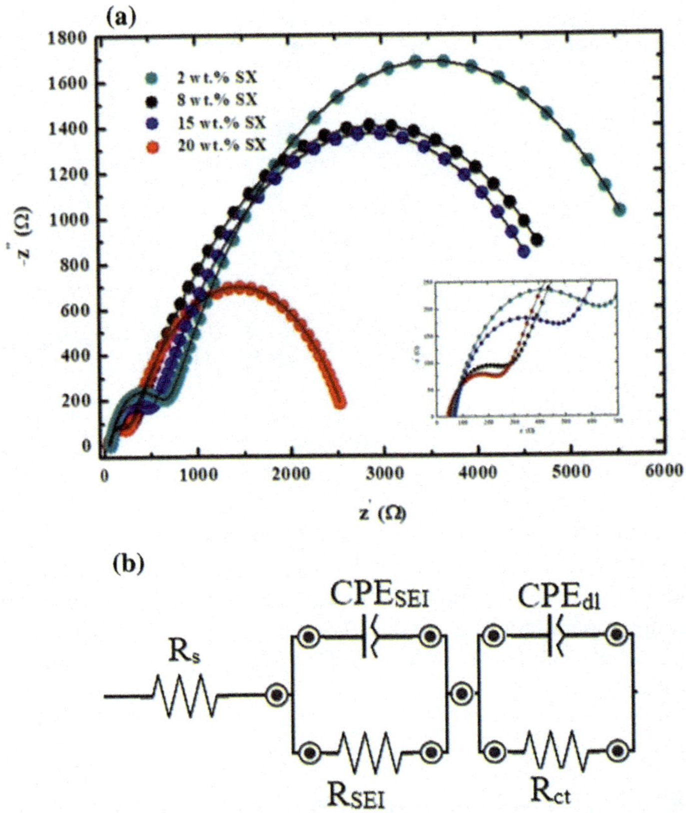

Fig. 2 **a** Nyquist plot of EIS and **b** equivalent circuit model of the fitted data for Al-air batteries using air cathode with different SX contents in 1 M KOH electrolyte

electrolyte. The plot consists of one small and one large semicircle at high and low-frequency regions, respectively. A starting point of the small semicircles in the Nyquist plot represents the equivalent series resistance R_s associated to the electrolyte solution, current collectors, and contact resistances [15, 16]. The small semicircle can be related with the solid electrolyte interphase layer resistance (R_{SEI}). R_{SEI} originates from the resistance of a layer that forms on the interface between the anode and electrolyte. The large semicircle can be associated with the charge transfer resistance of the cathode and electrolyte interface (R_{ct}). It can be seen from Table 1 that R_s increases slightly with the increase of the SX content from 2 to 20 wt.%. This increase is related to the addition of internal resistance arising from electrolyte drying in the separator, the contact between individual SX particles, and their contact with the current collector. When 20 wt.% SX is used for the battery, the smallest R_s is measured in the circuit. In the latter case, the SX content is sufficient to maintain the conductive particles together as a network and good connectivity are developed between SX particles in the composite layer. Therefore, they provide a facile electron

Table 1 Obtained fitting parameters for EIS data of Al-air batteries using air cathode with different SX contents in 1 M KOH electrolyte

Sample type	R_s (Ω)	R_{SEI} (Ω)	R_{ct} (Ω)	CPE_{SEI} $(\mu F s^{n-1})$	CPE_{ct} $(\mu F s^{n-1})$
2 wt.% SX	65.39	580	5800	6	180
8 wt.% SX	61.73	414	4800	10	260
15 wt.% SX	48.28	205	4600	18	300
20 wt.% SX	42.98	221	2500	24	360

pathway between the composite layer and current collector, leading to a fast reaction of electrolyte and oxygen in the pores. Then, for 15–2 wt.%, it enables the formation of reduced interconnection networks between SX particles in the composite layer because the blockage of large amount of electronic active layer by high content of graphite particles, and it results in a relatively large R_s compared to the case with 20 wt.%. The R_{SEI} decreases with increasing the SX content from 2 to 20 wt.%. As is well-known, SX is a nanostructural material with high porosity (about of 95%) and contains mainly micro and mesopores [17]. Thus, at higher content of SX, the composite layer on the air cathode has a high number of pore channels. In this case, more electrolyte ions can easily diffuse through the pores, and the reactions of oxygen and water can take place repeatedly to produce the higher amount of OH^- ions and reaction product $Al(OH)_3$ in the electrolyte medium. Since the surface of the Al anode has direct contact with the electrolyte and after the electrochemical reaction takes place, the product of this reaction accumulates on the surface and brings about corrosion. Moreover, the multiple repetitions of the electrochemical reactions lead to the accumulation of a significant amount of $Al(OH)_3$ on the anode surface. This results in an increase in the R_{SEI} which impedes OH^- migration to the Al anode in the discharge reaction. Such interpretation is in an agreement with the similar findings of the studies on lithium and magnesium-air batteries [5, 18]. The R_{ct} resistance is associated to the existence of an amount of the $Al(OH)_3$ reaction product [5] that enters the pores in the composite layer and is being adsorbed onto their surfaces. The charge transfer process between the pores surface in the composite layer and the electrolyte solution is being interrupted prematurely and leads to changes in the charge transfer resistance. The R_{ct} value increases significantly with increasing the SX content from 2 wt.% to 20 wt.%. This finding is attributed to the presence of micropores in SX. Micropores are too small to incorporate the $Al(OH)_3$, and according to literature [19], the entrance of micropores can easily be blocked by the $Al(OH)_3$; thus, these pores remain unfilled by oxygen and electrolyte ions. The increase in R_{ct} is attributed to the fact that the volume fraction of micropores increases when the SX content increases from 2 wt.% to 20 wt.%, so that the pore contact area blocked by the $Al(OH)_3$ increases, and therefore, the charge transfer resistance R_{ct} increases.

Figure 3 shows specific discharge capacity curves of the Al-air battery obtained using air cathodes with different SX contents in 1 M KOH electrolyte. It can be seen that the cell voltage of the battery having an air cathode layer with 20 wt.% SX

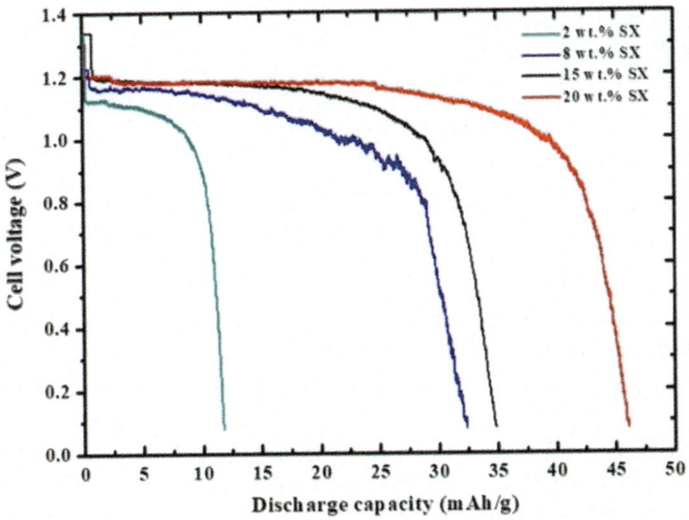

Fig. 3 Specific discharge capacity of Al-air battery using air cathodes with different SX contents in 1 M KOH electrolyte

amounts to approximately 1.2 V. When the content of SX decreases, the cell voltages declines as well. This decrease is attributed to an increase in the measured R_s as a result of the drying out of the separator [20]. It was also found that the specific discharge capacity increases with the increase of the SX content. Such behavior is related to a decrease in the measured R_{SEI} and R_{ct}, and this is in line with the interpretations from the literature [21]. It has been suggested that a reaction product layer of $Al(OH)_3$ is formed and grows on the surface of the Al anode and the entrance of micropores [22]. The existence of these layers reduces the catalytic activity in an electrochemical reaction on the Al anode surface and blocks the electrolyte ions from entering the micropores, thereby reducing the amount of OH^- ions produced in the micropores. As a result, such processes can lower the discharge capacity of the Al-air battery.

4 Conclusion

Porous silica xerogel derived from sago waste ash together with graphite has been successfully used as an active layer material for the air cathode of Al-air batteries. It was found that there is a clear correlation between the variations in SX/graphite composition and the related changes of the discharge capacity of the Al-air battery cells. The battery with an air cathode layer composition of 20 wt.% SX has demonstrated the highest discharge capacity of 46.6 mAh/g. Furthermore, the measured discharge capacity decreases from 46.6 to 12.3 mAh/g as the composition of SX decreases from 20 wt.% to 2 wt.%. A decrease in the discharge capacity is attributed to an increase of the resistance on the interface between the anode and electrolyte (R_{SEI}) and the charge transfer resistance of the cathode and electrolyte interface (R_{ct}) due to the accumulation of electrochemical reaction products on the anode surface and at the micropore entrances on the air cathode surface.

Acknowledgements This research was funded by the Ministry of Education, Culture, Research and Technology, (KEMENDIKBUD RISTEK), Republic of Indonesia, through the Project of the Penelitian Terapan Unggulan Perguruan Tinggi (PTUPT) in 2021. The authors would like to thank Mr. Nico and Mr. Hasyir for kindly helping to prepare the samples of the electrode material.

References

1. Liu Y, Sun Q, Li W, Adair K, Li J, Sun X (2017) Green Energy Environ 2:246–277
2. Fuentes RE, Colon-Mercado HR, Fox EB (2014) J Power Source 255:219–222
3. Han H, Jeon Y, Liu Z, Song T (2018) Appl Sci 8:209–2018
4. Zhao Y, Huang G, Zhang C, Peng C, Pan F (2018) Int J Electrochem Sci 13:8953–8959
5. Liew S, Juan J, Lai C, Pan G, Thomas C, Yang K, Lee T (2019) Ionics 25:1291–1301
6. Xue Y, Miano H, Sun S, Wang Q, Li S, Liu Z (2015) J Power Source 297:202–207
7. Fotouhi G, Ogier C, Kim J, Kim S, Cao G, Shen AQ, Kramlich J, Chung J (2016) J Micromech Microeng 26:055011–055019
8. Mori R (2016) J Electron Mat 45:3375–3381
9. Mainar A, Luis E, Colmenares C (2018) J. Energy Storage 15:304–328
10. Asghar H, Iqbal M, Iqbal M (2019) SN Appl. Sci 1:1396–1406
11. Cai X, Lai L, Lin J, Shen Z (2017) Mater. Horizons 4:945–976
12. Roev V, Ma S, Lee D, Im D (2014) J Electrochem Sci Technol 5:58–64
13. Obrovac M, Chevrier V (2014) Chem Rev 114:11444–11502
14. Aripin H, Mitsudo S, Sudiana IN, Tani S, Sako K, Fujii Y, Saito T, Idehara T, Sabchevski S (2011) J. Infrared Milli. Terahz Waves 32:867–876
15. Dees D, Gunen E, Prakash J, Abraham D, Jansen A (2005) J Electrochem Soc 152:A1409–A1417
16. Wang G, Yang L, Chen Y, Wang J, Bewlay S, Liu H (2005) Electro-chimica Acta 50:4649–5465
17. Duraes L, Ochoa M, Rocha N, Patricio R, Duarte N, Redondo V, Portugal A (2012) J Nanosci Nanotechnol 12:6828–6834
18. Kitaura H, Zhou H (2012) Adv Energy Mat 2:889–894
19. Mori R (2017) RSC Adv 7:6389–6395
20. Shen C, Xie J, Liu T, Zhang M, Andrei P, Dong L, Hendrickson M, Plichta J, Zheng P (2018) J Electrochem Soc 165:A2833–A2839
21. Cheng S, Zhang J, Liu H, Leng Y, Yuan A, Cao C (1998) J Power Sources 74:155–157
22. Neves S, Fonseca CP (2004) J Braz Chem Soc 15:395–399

Production of *Lactuca sativa L.* By Applying Household Waste Fertilizers

V. Soto-Aquino, C. Alvarez-Montalván, M. Baltazar-Ruíz, Y. Rojas-Castillo, R. I. Laredo-Cardenas, and J. C. Alvarez-Orellana

1 Introduction

The waste management is a common problem in most cities in the country, due to various factors such as population growth, the increasing amount of waste generated by poor education and poor environmental awareness of the population [1]. This is reflected in the inadequate practice they give to their household waste, destining them to dumps in some improvised cases, which leads to generate of vector proliferation foci that cause serious problems in human health. Likewise, those who carry out agricultural activities have phytosanitary problems, infertile soils that increasingly require high doses of fertilization, all this causes crops with high residuality, instability of microorganisms and alteration to the ecosystem and the environment.

The quality of an organic fertilizer is determined from its nutritional content and its ability to provide nutrients to a crop; this content is directly related to the concentrations of these nutrients in the materials used for their processing [2]. Composting has several advantages over incineration and landfilling and is an effective solution for recycling such waste. This is because it has lower operating costs, reduces environmental impacts, and most importantly, the final product can be used as fertilizer [3]. This technique not only reduces the amount of waste sent to landfills, but also contributes to social, ecological and economic improvement, being the best alternative for the management and transformation of organic waste.

V. Soto-Aquino
Universidad Nacional Intercultural de La Selva Central Juan Santos Atahualpa, Av. Perú 612, Pampa del Carmen, Chanchamayo, Perú

C. Alvarez-Montalván (✉)
Universidad Continental, Av. San Carlos, Huancayo, Perú
e-mail: calvarez@continental.edu.pe

M. Baltazar-Ruíz · Y. Rojas-Castillo · R. I. Laredo-Cardenas · J. C. Alvarez-Orellana
Universidad Nacional Del Centro Del Perú, Av. Marical Castilla, Huancayo, Perú

© The Author(s), under exclusive license to Springer Nature Singapore Pte Ltd. 2023
S.-M. Chen (ed.), *Proceedings of 10th International Conference on Chemical Science and Engineering*, Springer Proceedings in Materials 21,
https://doi.org/10.1007/978-981-19-4290-7_9

The lower content of macro- and micronutrients that make up both Biol and Bokashi is by their very nature that fundamentally their function is to act as activators of microfauna and microflora, respectively, which indicates that the value or power of organic fertilizer is not based on the content of traditional nutrients (N; P; K), but in the presence of microbial flora (bacteria and fungi). The bacterial flora is the one that will determine the health of the plant through the conservation of soil moisture, the formation of enzymes that will facilitate the effective nutrition of the roots, the conservation of soil pH and oxygenation of the soil [4].

According to the background, the objective of this work was to determine the efficiency of organic fertilizers from household waste, in order to comfort organic and sustainable agriculture, giving importance and value to organic household waste for its transformation into solid and liquid organic fertilizers [5]. For this reason, the present research work deals with the elaboration of liquid fertilizer (Biol) and solid fertilizer (Bokashi) from the conscientious segregation of each organic waste and at the same time, know the efficiency of these, in the performance of the *(Lactuca sativa L.)*.

2 Methodology

2.1 Place of Study

The present study was carried out in the area that is located in Peru, in the region of Huánuco, in the province of Marañón, in the district of Huacrachuco, in the Locality of Gran Vía, whose geographical position of its southern latitude is 8° 32 35″, its west longitude is 76° 11 28″ and is located at an altitude of 2936 m.a.s.l.m (see Fig. 1). According to the ecological map of Peru, established by the former National Office of Resource Assessment (ONERN), the place where the present experiment was executed is located in the natural life zone: tropical lower montane dry forest (bs-MBT) [6].

2.2 Sampling Method

The type of research was applied at an experimental level, through a design of random complete blocks (DBCA), with 03 treatments and 07 repetitions, making a total of 21 experimental units, experimenting in the cultivation of the species Lactuca sativa L. as shown in Fig. 2. In the present study, 02 types of organic fertilizers were elaborated (Biol and Bokashi) which were made with the collaboration of residents of the district of Huacrachuco from the collection of their organic household waste.

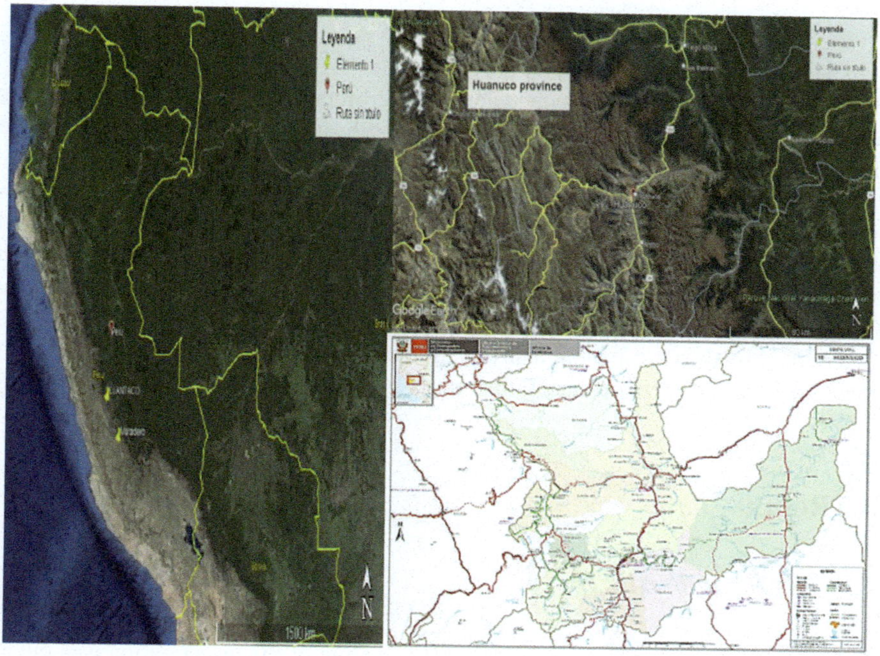

Fig. 1 Study location map

Fig. 2 Sampling process

2.3 Statistical Analysis

For the analysis of the data obtained and the interpretation of the results, the analysis of variance (ANVA) technique was used, and for the comparison of the averages, the Duncan significance test was applied at the 0.05 and 0.01 probability levels.

When recording head diameter data, the width and length of the lettuce leaves were determined using a graduated ruler, and to measure the weight, a graduated balance was used. The analysis of the chemical composition of Biol and Bokashi was carried out in the Soil Laboratory of the National Agrarian University of the Jungle. The economic evaluation in the elaboration of the Biol and the Bokashi was determined by the cost–benefit analysis; in the same way, the environmental impact assessment was determined using the matrix method.

3 Results

3.1 Organic Fertilizers

The collection of organic household waste by nature of the investigation was intentionally based on 34 families, of which the most generated waste that was manure, amounting to 251.68 kg; in the case of plant remains 136.56 kg and 23.90 kg when referring to the amount of ash. The results in costs both in supplies, materials, tools, transportation of materials and unskilled labor, it rises with a total amount of 1353.75 soles. The fertilizers obtained according to Table 1 present a different pH, where the BIOL seems higher due to its high concentration of phosphorus and potassium, whereas Bokashi is slightly acidic due to its higher abundance of nitrogen and magnesium. As for calcium, it seems to have of little relevance in the composition of both fertilizers.

Table 1 Chemical characterization of organic fertilizers

Type of fertilizer	*pH*	Nitrogen *(N)*	Potassium *(K)*	Magnesium *(Mg)*	Calcium *(Ca)*	Phosphorus *(P)*
Biol	Slightly alkaline (7.8)	0.15%	2.34%	0.61%	2.74%	0.94%
Bokashi	Slightly acidic (6.44)	0.84%	0.26%	1.11%	0.47%	0.39%

3.2 *Lettuce Production*

In Table 2, we can notice that for the number of leaves per plant at harvest time, high statistical significance is for the source of treatments in the F test. The analyses show a correct application of the experiment, avoiding the effect of edges and middle of the plots. Where the number of leaves, diameter and weight is the variables with the greatest significance, this suggests a significant difference between the Bokashi and Biol treatments.

In Table 3, Duncan's test of significance at the 0.05 probability level shows us that the first place is always for Bokashi, it even far exceeds Biol, which is in second place. On the other hand, the number of leaves there is a marked difference but only not very wide, which shows that both fertilizers were effective for the production of lettuce. In the case of diameter, Bokashi far exceeds Biol, which may be due to a higher concentration of nitrogen in the Bokashi; similarly, the weight also shows a significant difference, where Bokashi shows a weight of 415 gr, demonstrating greater effectiveness with the rest of the treatments for the production of lettuce.

Table 2 Analysis of variance of the metric characteristics of lettuces according to the types of fertilizers

Variables	F value calculated	F value tabulated		Sig (0.99)
		0.05	0.01	
Number of leaves	241.47	3.89	6.93	**
Overall diameter(cm)	159.38	3.89	6.93	**
Weight (gr.)	459.09	3.89	6.93	**
Number of non-commercial leaves	14.70	3.89	6.93	**
Leaves width	77.49	3.89	6.93	**
Leaves length	36.57	3.89	6.93	**

Table 3 Duncan's significance test for number of leaves at harvest time

Variable	Ranking	Mean	Organic fertilizers	Significance level			
				0.05		*0.01*	
Number of leaves	1	34.14	Bokashi	*a*		*a*	
	2	32.14	Biol		*b*	*a*	
	3	17.37	No treatment		*c*	*c*	
Overall diameter *(cm)*	1	35	Bokashi	*a*		*a*	
	2	29	Biol		*b*	*b*	
	3	21	No treatment		*c*		*c*
Total weight *(gr)*	1	415.71	Bokashi	*a*		*a*	
	2	345	Biol		*b*	*b*	
	3	295	No treatment		*c*	*c*	

4 Discussion

Having carried out the analysis of the chemical composition, in the case of Bokashi with respect to nitrogen content, it amounted to 0.84%, with respect to Biol which only contained 0.15%, and after evaluation in lettuce cultivation with respect to head diameter, number of leaves and head weight, the treatment with Bokashi showed high statistical significance with respect to Biol treatment. This indicates that it is the fundamental component of all organic molecules involved in the processes of plant growth and development [7].

The economic evaluation is favorable because having produced 500 kg of Bokashi whose sale price per kilo is 0.70 cents projecting only 04 campaigns in year zero has s/. 1400.00 and in the case of Biol whose production was 200 L whose sale price at s/. 6.00, projecting only 04 campaigns per year will have the sum of s/. 4 800. 00, totaling the benefits to s/. 6200.00 with a production cost of s/. 8 122.50, this indicates that the positive benefits will be given from year 1 onward [8].

In the same way, the composting generated by the homes of the different families helps to designate in a more orderly and favorable way for the environment without wasting the nutrients they provide [9]. The use of this waste as an efficient means of rational recycling of nutrients, through its transformation into organic fertilizers, helps the growth of plants and contributes to improving or maintaining many soil properties. The benefits of using organic amendments such as Bokashi are widely known worldwide [10].

The demands in international markets require the observation and certification of an organic production, which goes to the benefit of the same environment in view of the deterioration of the environment, pollution, global warming and other environmental problems that have arisen over the years; therefore, in transnational companies today, there is a demand for organic production for the benefit of human health [3].

Regarding the nutrients provided by composting, the contents of Ca, K and N were those with the highest concentration in the compost, followed by Fe, Zn, Mn, Cu and B, in smaller proportions. According to these authors, domestic organic solid waste is an important source of organic matter, providing essential nutrients for plants, which may be available when composting is added to the soil [8].

5 Conclusions

The chemical composition of the liquid fertilizer given in percentage is: 0.15% of nitrogen, 0.94% of phosphorus, 2.34% of potassium, 0.61% of magnesium, 0.25% of sodium, 2.74% of calcium with a pH of 7.85. In the case of solid fertilizer, its chemical composed of: 0.84% of nitrogen, 0.39% of phosphorus, 0.26% of potassium, 0.11% of magnesium, 0.02% of sodium, 0.47% calcium and a pH of 6.44.

Having performed the F test with respect to the number of leaves, head diameter, weight at harvest time, leaf width and leaf length, it is observed that the statistical difference between the treatments is highly significant, finding no significance for the source of variation with respect to repetitions.

From Duncan's significance test at the 0.05 and 0.01% probability level, Bokashi's treatment ranked first, statistically surpassing in the order of merit from second to third place.

About 500 kg of solid fertilizer and 200 L of liquid fertilizer were produced with a total amount of 1353.75 Nuevo's soles; carrying out the economic evaluation in year 0, four campaigns would be produced and from year 1 to year 5 six campaigns for each year.

When carrying out the environmental impact assessment in the production of organic fertilizers, it is concluded that it does not generate a negative impact; on the contrary, it allows the proper usage of organic household waste.

References

1. Krounbi L, Enders A, van Es H, Woolf D (2019) van Herzen B y Lehmann J Biological and thermochemical conversion of human solid waste to soil amendments. Waste Manag 89:366–378
2. El-Mogy MM, Abdelaziz SM, Mahmoud AWM, Elsayed TR, Abdel-Kader NH, Mohamed MIA (2020) Comparative effects of different organic and inorganic fertilisers on soil fertility plant growth, soil microbial community, and storage ability of lettuce. Agriculture 66:87–107
3. Chojnacka K, Moustakas K, Witek-Krowiak A (2020) Bio-based fertilizers: A practical approach towards circular economy. Bioresource Technol 295:122223
4. Santos Viana J dos, Roque Borda CA, Palaretti LF (2020) Application of bokashi organic fertilizer in production oflettuce (Lactuca sativa) Horticulture Int J 4:200–1
5. Sang-mo K, Arjun A, Dibya B, Ho-jun G, Gim M, Joon-ik S, Jin Y S, In-Jung L (2022) Comparison of effects of chemical and food waste-derived fertilizers on the growth and nutrient content of lettuce
6. Holdridge LR, Tosi JA (1967) Life zone ecology (San Jose, Costa Rica: Tropical Sceince Center)
7. Sigurnjak I, Brienza C, Snauwaert E, De Dobbelaere A, De Mey J, Vaneeckhaute C, Michels E, Schoumans O, Adani F, Meers E (2019) Production and performance of bio-based mineral fertilizers from agricultural waste using ammonia (stripping-)scrubbing technology. Waste Manag 89:265–74
8. Natsheh B y Abu-Khalaf N 2020 Influence of different types of fertilizers application on the lettuce (Lactuca sativa L.) growth and quality *Palestine Technical University Kadoorie Research Journal* **8** 40–53
9. Abunde Neba F, Asiedu NY, Addo A, Seidu R (2020) Attainable regions and fuzzy multi-criteria decisions: modeling a novel configuration of methane bioreactor using experimental limits of operation Bioresource Technol 295:122273
10. Moraes VH, Giongo PR, Albert AM, Arantes BHT, Mesquita M (2021) Development of lettuce varieties in different organic wastes as substrate. Revista Facultad Nacional de Agronomia Medellin 74:9483–9

Molecular Docking Studies on the Binding Affinity of Alpha-Conotoxins on Voltage-Gated Sodium Ion Channel Using an Incremental Genetic Algorithm Approach

L. L. Tayo, A. C. Aquino, and E. C. Pasamba

1 Introduction

Cone snails are attractive therapeutic targets due to their naturally occurring neuroactive peptides also known as conotoxins or conopeptides found in their venom [1]. Upon injection of their venom to the target organism, severe effects such as change in behavior, uncontrollable shaking, convulsions, and ultimately paralysis are observed in the prey [2]. Great diversity of conotoxins is observed due to equally diverse populations of the cone snails based on size, shape, and patterns [2]. Most importantly, hundreds of conopeptides are present in the venom of one cone snail much similar to a cocktail of different toxic peptides [2]. This mixture both enhances the damage to the prey and poses a challenge to researchers in identifying such proteins [2].

The molecular diversity of conotoxins has great importance in the fields of structural biochemistry and mammalian neuromuscular systems [2–7]. As reported by Tayo et al., the high variability and difficulty in defining uniformity of conotoxins may have originated from the difference in expression of conotoxins along the venom duct of a Conus organism. Environmental factors also contribute to the variable development of these venomous apparatus [3] which could result to similarly diverse cocktails of neurotoxic peptides [4]. Hence, further studies on how certain conotoxin classes among highly diverse conotoxins act on specific nociceptors are urgently needed [5]. Voltage-gated sodium channels (Na_v) are one of the most important channels in the somatosensory nervous system as they amplify the signal to be transmitted to the spinal cord and recognized by the brain [6]. Sodium-gated channels

L. L. Tayo (✉) · A. C. Aquino · E. C. Pasamba
School of Chemical, Biological and Materials Engineering and Sciences, Mapua University, Intramuros, Manila, Philippines
e-mail: lltayo@mapua.edu.ph

© The Author(s), under exclusive license to Springer Nature Singapore Pte Ltd. 2023

S.-M. Chen (ed.), *Proceedings of 10th International Conference on Chemical Science and Engineering*, Springer Proceedings in Materials 21,
https://doi.org/10.1007/978-981-19-4290-7_10

may be blocked, movement may be limited, and ultimately activation may be inhibited resulting to subtle taser-like effects, hyperactivity of the nervous system, to more severe effects such as irreversible paralysis [1]. The importance of conotoxins on pain management has already been established [7], and several drugs have already been formulated toward pain relief [1] but has not yet been studied for their interactions with voltage-gated sodium channels.

In this study, α-conotoxins from the A superfamily with cysteine frameworks I taken from ConoServer (conoserver.org) as ligands and an open-channel conformation of a voltage-gated sodium channel from the marine bacterium Magnetococcus sp. (strain MC-1) named NavMs is used as receptor. Then, incremental molecular docking using AutoDock Vina in DINC 2.0 (dinc.kavrilab.org) is performed. To predict the best α-conotoxin to be used as potential drug design template, the binding energies were used to determine the binding affinities of the said class to voltage-gated sodium channel. Results from this study provide further insights into the potential druggability of α-conotoxins and also the mechanism of this specific class of conotoxins with voltage-gated sodium channels.

2 Research Methodology

2.1 Data Gathering

The ligands in the study are alpha-conotoxins from the A superfamily with cysteine framework I. Tertiary structures are taken from ConoServer (conoserver.org). A total of 19 conotoxins is docked with the receptor. The receptor used in this study is the open-channel conformation of a voltage-gated sodium channel from the marine bacterium Magnetococcus sp. (strain MC-1) (Na_vMs) as reported by previous study [8]. At the time of this writing, the said voltage-gated sodium channel is the only solved structure that contains the complete domains of this channel [9]; therefore, its pore structure which has been elucidated for its opening and closing mechanism is chosen for this study [8].

2.2 Molecular Docking

Proteins are cleaned from different conformations, incomplete residues, complexation with bound receptors, and water molecules; all cleaning processes were performed in PyMOL; specifically, the clean and extract object molecule were used. The DINC 2.0 (dinc.kavrilab.org) is used for all molecular docking of ligands to receptor. DINC is a parallelized meta-docking method for the incremental docking of large ligands using AutoDock Vina. The box center is chosen to be the ligand center, and the box dimensions are set to be based on the ligand. The ligand-receptor complex is viewed with the option "center for ligand" in all complexes. It utilizes a heuristic approach toward determining the best solution among different binding conformations of proteins [10]. Briefly, DINC uses AutoDock's scoring function wherein a lower score corresponds to a higher ranking in conformation prediction and docking accuracy is computed through the root mean square deviation (RMSD) between the resulting conformation of the ligand compared to the ligand conformation obtained from PDB [11]. Hence, the lowest energy conformation pertains to the docked pose with the lowest score, whereas the energy scores from the three largest clusters of the docked conformations were also obtained.

2.3 *Results Analysis*

The binding energies in kcal/mol and RMSD values were obtained for each conotoxin voltage-gated channel complex to infer the binding affinity of different alpha-conotoxins with Na_vMs.

3 Results

Knowledge on protein families targeted by conotoxins has already been discussed by a previous study [12]. Molecular basis of conotoxin binding with receptors, such as binding affinity and RMSD, provides quantitative insights both in classification and function of target receptors with conotoxins [12]. Computational studies on the binding affinity of α-conotoxins with voltage-gated sodium channels are few; hence, there is an urgent need to understand the behavior of α-conotoxins as a class with a highly important component in the nervous system. In this study, genetic algorithm along with an incremental approach toward docking was utilized to obtain the best solution for the two proteins.

A protein–ligand complex is understood as a solute in a solvent which participates in interactions involving heat transfer [13] such as signal transduction [5]; hence, the laws of thermodynamics may be applied here [13]. Briefly, the Gibb's free energy change, ΔG, is the relative change of free energy values as one molecule conformation is changed into another conformation during biological conditions [14]. The most negative value indicates the stability of the interactions between the ligand and the receptor which corresponds to the best conformation of the docked complex [13].

Results showed that the range of binding energies between Na_vMs and α-conotoxins is between -4.6 ± 0.0000 to -6.87 ± 0.2357 kcal/mol (Table 1). Also, a close range of binding energies is observed and may also be characteristic only to α-conotoxins. DINC has two approaches of determining the best solution specifically through ranking by energy or clustering by RMSD (Table 1). Based on both approaches, α-conotoxin binding with Na_vMs is a favorable and spontaneous process thermodynamically as observed from the very negative values of binding energies.

Table 1 Thermodynamic stability of α-conotoxins with NavM

		Lowest energy conformation		Lowest energy from the three largest clusters	
No.	PDB ID	Binding energy, kcal/mol	RMSD	Binding energy, kcal/mol	RMSD
1	1HJE	−4.6 ± 0.0000	12.00 ± 0.3112	−4.23 ± 0.1247	11.83 ± 0.0492
2	1B45	−5.3 ± 0.0816	11.24 ± 0.2604	−5.00 ± 0.2160	11.59 ± 0.4926
3	1DG2	−6.00 ± 0.0816	9.39 ± 0.0205	−5.43 ± 0.4109	10.76 ± 0.9543
4	1IM1	−6.23 ± 0.0471	10.06 ± 2.7362	−5.47 ± 0.3299	10.16 ± 2.6729
5	1IMI	−6.77 ± 0.0943	12.05 ± 0.8957	−6.30 ± 0.6480	11.65 ± 1.0321
6	1M2C	−5.27 ± 0.0943	11.58 ± 0.0873	−4.40 ± 0.2943	12.40 ± 0.7476
7	1MTQ	−4.77 ± 0.1700	11.72 ± 0.7299	−4.30 ± 0.1632	11.61 ± 0.8387
8	1MXN	−6.8 ± 0.0000	9.86 ± 0.0047	−5.17 ± 1.1897	10.32 ± 0.6155
9	1NOT	−5.13 ± 0.0471	12.88 ± 0.4411	−4.30 ± 0.5715	13.73 ± 0.9080
10	1QMW	−5.5 ± 0.0000	11.11 ± 0.1485	−4.73 ± 0.4642	11.30 ± 0.5739
11	1UL2	−5.53 ± 0.0943	11.62 ± 1.3081	−5.27 ± 0.2494	10.68 ± 0.6107
12	1XGB	−6.2 ± 0.0816	10.00 ± 0.0638	−5.53 ± 0.5312	9.67 ± 0.3817
13	2I28	−6.87 ± 0.2357	11.44 ± 0.0450	−6.10 ± 0.4242	10.95 ± 0.7376
14	2GCZ	−5.97 ± 0.0471	13.31 ± 0.0455	−5.70 ± 0.4242	13.16 ± 0.2491
15	2NS3	−6.87 ± 0.0471	11.72 ± 1.3678	−6.00 ± 0.1632	11.37 ± 1.8238
16	2JUT	−5.07 ± 0.1700	11.55 ± 0.4340	−5.03 ± 0.2054	10.59 ± 1.0335
17	2C9T	−3.77 ± 0.0471	10.07 ± 0.0572	−3.20 ± 0.4242	10.91 ± 1.5344
18	4EZ1	−5.97 ± 0.0471	12.90 ± 0.0189	−4.97 ± 0.4784	12.44 ± 0.5629
19	2BYP	−4.77 ± 0.1700	3.25 ±4.0821	−3.43 ± 0.7930	4.45 ± 2.9322

Table 1 also showed that conotoxin BuIA with PDB ID 2I28 is the most probable active α-conotoxin since it acquires a binding energy value of −6.87 ± 0.2357 kcal when docked with Na_vMs. In terms of structural conformation, the scoring by RMSD values is taken into account. Results showed that the RMSD values are very large which mean that the resulting conformation bound to Na_vMs receptor is highly different from the true conformation before binding.

4 Discussion

The importance of conotoxins on pain management has already been established [7]; however, few drugs have been formulated toward pain relief [1, 3]. As reported by Gao et al., research into drug patents and pre-clinical trials from 1998 to 2017 has determined that one conotoxin-derived drug which targets sodium channels has undergone pre-clinical trial for neuropathic pain treatment before it was terminated.

The conotoxin MrVIB as a novel drug has the ability to distinguish different sodium channels by exhibiting specific affinities and reduced pain in animal model at low dosage only [15]. Additionally, only δ, μ, μO, and ί-conotoxins only were discovered to modulate sodium channels [16]. Therefore, studies on conotoxin behavior with other important pharmacological targets are needed to understand its mode of action and future applications in drug design and pain relief treatments.

In this study, the mode of action of α-conotoxins to voltage-gated sodium channels was predicted. Thermodynamically favorable and spontaneous binding with NavMs was observed among α-conotoxins, and the resulting protein–protein complexes using an incremental docking approach based on genetic algorithm exhibited a highly different conformation from its native state. Several factors contribute to a high RMSD value such as peptide and receptor flexibility [10], effects of desolvation [17], identification of binding site [17], and other challenges being addressed by the chosen algorithm [13]. In another study by Gomez et al., BuIA is predicted to belong in Group 1 α-conotoxins specifically in subgroup 1 based on its structural alignment. Specifically, conotoxins of this class were classified based on pairwise alignment and root mean square deviation (RMSD) values [18]. Generally, low RMSD values were obtained for this class which established that sequence variations may be used to infer conotoxin relationships. In terms of protein structures, results in this study have determined characteristic binding affinities of conotoxins with a voltage-gated sodium channel which is an equally important receptor in the nervous system as nAChRs. The structure of conotoxin BuIA (PDB ID 2I28) was also utilized as a template to search for highly similar analogs that target human nAChRs for the purpose of potential analgesics in another study [19]. Briefly, molecular docking was performed on the said conotoxin with different subunits of nAChRs, and results showed high affinity with specific subunits; hence, BuIA was selected for further experiments [19]. The inhibitory effect of conotoxin GIC (PDB ID 1UL2) was also observed with a specific type of nAChRs [20], and key interacting portions of this conotoxin have been elucidated computationally [21]. However, it has been reported that a novel α-conotoxin Pl168 does not display fluorescent activity in cell lines with VGSCs along with nAChRs and voltage-gated calcium channels; hence, further studies on the behavior of α-conotoxins with various receptors still constitute an active part of research [22]. Previous studies have shown the importance of molecular docking in the identification of structural variations that may enhance their analgesic properties toward important neuroreceptors [12]. Therefore, further studies on the molecular interactions of conotoxins and the basis of conformation change for additional receptors are recommended to understand both the evolution of conotoxin binding and its mode of action which may be expedited through a computational approach.

5 Conclusion

Previous studies showed that α-conotoxins recognize nAChRs only. In this study, the binding between the bacterial voltage-gated sodium channel and α-conotoxins was possible using incremental docking software based on genetic algorithm. Binding occurs as a favorable spontaneous process thermodynamically as observed from low energy scores and the close range of binding energies obtained which may be characteristic only to α-conotoxins. Additionally, the resulting docked complexes were highly different from the true ligand conformation based on RMSD values. Hence, further studies toward other factors that may contribute to this are recommended. Molecular docking studies of α-conotoxins on voltage-gated sodium channels give insights on potential novel α-conotoxins as analgesic drugs to neuroreceptors different from the widely accepted nAChRs.

Acknowledgements The authors wish to express their most heartfelt gratitude to Dr. Lemmuel L. Tayo for the guidance and expertise on conotoxins throughout this study.

References

1. Green BR, Olivera BM (2016) Venom peptides from cone snails: pharmacological probes for voltage-gated sodium channels. Curr Top Membr 78:65–86
2. Puillandre N, Dutertre S (2018) The conoidea and their toxins: evolution of a hyper-diversified group. Biodivers Evol 227–49
3. Gao B, Peng C, Yang J, Yi Y, Zhang J, Shi Q (2017) Cone snails: a big store of conotoxins for novel drug discovery. Toxins (Basel) 9(12):1–17
4. Tayo LL, Lu B, Cruz LJ, Yates JR (2010) Proteomic analysis provides insights on venom processing in conus textile. J Proteome Res 9(5):2292–2301
5. Jin AH, Muttenthaler M, Dutertre S, Himaya SWA, Kaas Q, Craik DJ, Lewis RJ, Alewood PF (2019) [*Conotoxins: Chemistry and Biology*]. Chem Revi. Am Chem Soc 119:11510–11549
6. Bennett DL, Clark XAJ, Huang J, Waxman SG, Dib-Hajj SD (2019) The role of voltage-gated sodium channels in pain signaling. Physiol Rev 99(2):1079–1151
7. Munasinghe NR, Christie MJ (2015) Conotoxins that could provide analgesia through voltage gated sodium channel inhibition. Toxins (Basel) 7(12):5386–5407
8. McCusker EC, Bagnéris C, Naylor CE, Cole AR, D'Avanzo N, Nichols CG, Wallace BC (2012) Structure of a bacterial voltage-gated sodium channel pore reveals mechanisms of opening and closing. Nat Commun 3:1–8
9. Sula A, Booker J, Ng LCT, Naylor CE, Decaen PG, Wallace BA (2017) The complete structure of an activated open sodium channel. Nat Commun 8:2–10
10. Antunes DA, Devaurs D, Moll M, Lizée G, Kavraki LE (2018) General prediction of peptide-mhc binding modes using incremental docking: a proof of concept. Sci Rep 8(1):1–13
11. Dhanik A, McMurray JS and Kavraki LE. DINC (2013) A new autodock-based protocol for docking large ligands. BMC Struct Biol 13(SUPPL.1):1–14
12. Mansbach RA, Travers T, Fair JM, Gnanakaran S (2019) Snails in silico: a review of computational studies on the conopeptides. Mar Drugs 17(3):1–34
13. Du X, Li Y, Xia YL, Ai SM, Liang J, Sang P, Ji XL, Liu SQ (2016) Insights into protein-ligand interactions: mechanisms, models, and methods. Int J Mol Sci 17(2):1–34
14. [Kukol A (2014) [*Molecular modeling of proteins: Second edition*]. Vol 1215, Molecular Modeling of Proteins: Second Edition, pp 1–474
15. Ekberg J et al (2006) MO-conotoxin MrVIB selectively blocks NaV1.8 sensory neuron specific sodium channels and chronic pain behavior without motor deficits. Proc Natl Acad Sci US A 103(45) 17030–17035
16. Gallo A, Boni R, Tosti E (2020) Neurobiological activity of conotoxins via sodium channel modulation. Toxicon 187(July):47–56

17. Ferreira LG, Dos Santos RN, Oliva G, Andricopulo AD (2015) [*Molecular docking and structure based drug design strategies*]. Vol 20, Molecules, pp. 13384–13421
18. Gomez MC, Aquino AMC, Matira AR, Alvarico RAD, Valbuena RE and Tayo LL (2019) Alpha family of conotoxins: an analysis of structural determinants. ACM Int Conf Proceeding Ser 40–6
19. Liu C et al (2019) Rationally designed α-conotoxin analogues maintained analgesia activity and weakened side effects. Molecules 24(2):1–15
20. Ho TNT, Abraham N, Lewis RJ (2020) Structure-function of neuronal nicotinic acetylcholine receptor inhibitors derived from natural toxins. Front Neurosci 14(November):123
21. Lee C, Lee SH, Kim DH, Han KH (2012) Molecular docking study on the A3β2 neuronal nicotinic acetylcholine receptor complexed with α-conotoxin GIC. BMB Rep 45(5):275–280
22. Wilson DT, Bansal PS, Carter DA, Vetter I, Nicke A, Dutertre S, Daly NL (2020) Characterisation of a novel A-superfamily conotoxin. Biomedicines. 8(5):2–11

Determination of L-Citrulline Content in the Mesocarp of the Verde, Pintón and Maduro Fruit of Citrullus Lanatus (Watermelon) in Pucallpa

C. Ruiz, J. Estela, S. Camargo, Da A. Cruz, E. Daza, S. Zavala, and N. Balbin

1 Introduction

In the region of Ucayali, the watermelon fruit *Citrullus lanatus* is marketed as fresh fruit, of which the part that is mostly consumed is the pulp, however, according to Durán in 2014, mentions that the watermelon peel is a carrier of an amino acid called citrulline, which is beneficial to humans as it is a precursor for the synthesis of arginine in our body [1]. In this department there are areas devoted to the planting of watermelon, among which are the districts of Manantay, Yarinacocha, Campo Verde, Aguaytia and others. During the sale and marketing of watermelon in the department of Ucayali it is observed that the watermelon peel is discarded without giving an alternative use, contributing to environmental pollution by organic solid waste. In Ucayali, watermelon production in 2019 was 612 metric tons, with a sowing of 200 hectares and a harvested area of 25 ha [2].

The objective of the present investigation was to determine the citrulline content present in the mesocarp of watermelon produced in Ucayali at different stages of fruit ripening. Research conducted in the USA by Rimando and Perkins analyzed the citrulline content in watermelon mesocarp of different varieties by HPLC, obtaining results between 1.2 mg/g and 3.1 mg/g (fresh weight analysis) in red-fleshed watermelon with seeds [3]. The results of the present work showed that the highest content of citrulline was presented in the mesocarp of maduro fruit, with 1.05 mg/g (fresh

C. Ruiz (✉) · J. Estela
Universidad Nacional de Ucayali, Pucallpa, Peru
e-mail: carlos_ruiz@unu.edu.pe

S. Camargo · N. Balbin
Universidad Continental, Huancayo, Peru
e-mail: scamargo@continental.edu.pe

D. A. Cruz · E. Daza · S. Zavala
Universidad Nacional Agraria de La Selva, Tingo Maria, Peru

© The Author(s), under exclusive license to Springer Nature Singapore Pte Ltd. 2023
S.-M. Chen (ed.), *Proceedings of 10th International Conference on Chemical Science and Engineering*, Springer Proceedings in Materials 21,
https://doi.org/10.1007/978-981-19-4290-7_11

weight analysis); optimal value close and slightly lower than the research done by Rimando and Perkins in USA; the results of this research will allow to consider alternatives for the adequate use of watermelon mesocarp.

2 Literature Review

2.1 *General*

According to Rimando and Perkins in 2005, in a study called "Determination of citrulline in watermelon rind" it is concluded that the citrulline content present in the mesocarp of red-fleshed seeded watermelons ranges from 1.2 mg/g to 3.1 mg/g [3].

According to Collins, L-arginine works primarily through increasing nitric oxide levels in the blood. Higher levels of nitric oxide relax and dilate blood vessels, which increases blood circulation to various organs including the heart, skin, and sexual organs [4].

2.2 *Characteristics of* **Citrullus Lanatus** *(Watermelon)*

According to Giaconi, it is a monoecious herbaceous plant whose origin is presumed to be in Africa, where it still grows wild [5].

Development. According to Peñaloza, the germination of the seeds, which are distributed by the pulp, requires relatively high temperatures, minimum of 12–16 °C with an optimum between 28 to 35 °C. They are flattened and generally of length less than twice the width, ovoid, hard, of variable weight and color (white, brown, black, yellow, yellow, mottled). The appearance of the radicle is limited by low temperatures [6].

Chemical composition of watermelon. Watermelon is a fruit with a water content of 91–93%, with small amounts of proteins, fats, minerals, and vitamins, and it is a fruit poor in phenolic compounds and vitamins but contains an important source of lycopene and L-citrulline [4] but contains an important source of lycopene and L-citrulline [3].

It is a fruit with healthy properties, and the richness in lycopene of watermelon makes it a worthy rival of tomato in the prevention of prostate cancer and against cellular aging; also, scientific studies have confirmed that watermelon stands out in the content of citrulline, an amino acid important for the synthesis of arginine, which acts as a dilator of blood vessels [7].

2.3 Citrulline

A non-essential amino acid first identified from the juice of watermelon *Citrullus vulgaris* and subsequently obtained from tryptic digestion of casein, citrulline has also been isolated from other cucurbit fruits such as melon, cucumber, and squash. In most mammals, the small intestine is the main source of circulating citrulline and is used for the synthesis of arginine [3].

Citrulline synthesis in the human body. Citrulline is synthesized almost exclusively by the small intestine, and it is for this reason that plasma citrulline has been identified as a biomarker of functional small intestinal enterocyte mass [8].

Intestinal Biosynthesis. The enzymatic pathway of intestinal citrulline biosynthesis is a complex sequence of five mitochondrial enzymes in enterocytes. The key enzymes are ornithine aminotransferase (OAT) and pyrroline-5-carboxylate synthase (P5C); the latter of which is expressed alone in the intestinal mucosa [9].

Hepatic Biosynthesis. Citrulline can be synthesized directly in liver tissue from the urea cycle, but due to the high activity of the enzyme arginine succinate synthase (ASS), the citrulline produced is rapidly converted to arginosuccinate [10].

2.4 High-performance Liquid Chromatography with Diode Array Detector (HPLC–DAD)

This type of chromatography makes use of diodes that allow the conversion of the intensity of the emitted UV light into an electrical signal; state-of-the-art equipment such as the Agilent 1200 series chromatograph allows a higher resolution since it has a column with different configurations for different particle sizes to be analyzed, as well as length, diameter, among others, which allows the analyst to customize the separation of the amino acid according to this specifications [11].

3 Materials

3.1 Place of Execution

The present work was carried out in the following places: The watermelon was collected in the district of Nueva Requena, located in the province of Coronel Portillo, department of Ucayali with coordinates: latitude of −8.3341°, longitude of −74.562° and altitude of 153 m.a.s.l. [12]. This information can be found in Fig. 1

Fig. 1 Location map

3.2 Materials

Watermelon mesocarp *Citrullus lanatus*, water, ice cubes, sodium hypochlorite (5%), alcohol (96°), hydrochloric acid (1 N), acetonitrile, methanol, ultrapure water (HPLC water), chillo, spoon, plastic tubs, stainless steel bowls, chopping board, sponges, absorbent cloths, thermally insulated cooler, polyethylene bags with hermetic seal, indelible fine tip marker, 15 ml plastic measuring cup, protective accessories (apron, gloves, mask, cap).

Teams. Camry digital weighing balance, BDF 86V100 laboratory freezer with a temperature range of −86 °C to −10 °C, Labtron refrigeration equipment with a temperature range of −4 °C to 10 °C, Agilent 1200 series HPLC chromatographic equipment with Zorbax Eclipse AAA Rapid Resolution column and diode array detector.

4 Methods

4.1 Methodology

Choice of raw material. *Citrullus lanatus* (watermelon) was selected considering aspects such as fruit size and ripening stage.

Reception of raw material. It was considered that the watermelon is free of physical damage such as bumps, cuts, etc., as this would affect the quality of the sample.

Sample Preparation. The steps for sample preparation of 100 g of watermelon mesocarp at three stages of ripening are described below:

4.2 *Statistical Research Design*

Mathematical Model

$$Y_{\mathrm{ij}} = \mu + \tau_i + \varepsilon_i \tag{1}$$

where

Y_{ij}: Outcome of the "*i*" subject under the "*i*" treatment.
μ: Common mean of all the data of the experiment.
τ_i: Effect of "*i*" treatment.
ε_i: Experimental error or random sampling effect.

Analysis of Variance DCA. A treatment number was assigned for each ripening stage of watermelon mesocarp as found in Table 1.

The number of repetitions was assigned to each of the treatments under study as found in Table 2.

Data analysis. Statistical tests were performed: DCA test and Tukey's multiple comparison of means test at a significance level of 5%, to determine the degree of variability between the treatments under study. The statistical software Statgraphics Centurion was used for the treatment of the data; in addition, the presentation and graphs of data were made in Microsoft office.

Table 1 Treatments under study

Treatments	Mesocarp condition
T1	Verde
T2	Pintón
T3	Maduro

Table 2 Assignment of repetitions

Treatments	Repetitions		
	1	2	3
T1	R1	R2	R3
T2	R1	R2	R3
T3	R1	R2	R3

Table 3 Citrulline content present in 100 g of watermelon mesocarp

Sample	R1	R2	R3	Average (mg/100 g)
Verde	38.333	38.37	38.377	38.36
Pintón	46.991	46.912	47.002	46.968
Maduro	105.567	105.611	105.631	05.603

5 Results

5.1 Watermelon Mesocarp L-Citrulline Content

The citrulline content of each sample is analyzed, obtaining an average at each stage of maturation. The number of replicates was assigned to each of the treatments under study as found in Table 3.

5.2 Citrulline Kinetics in HPLC Analysis

The time of retention of citrulline present in the mesocarp of maduro watermelon is found in Fig. 2.

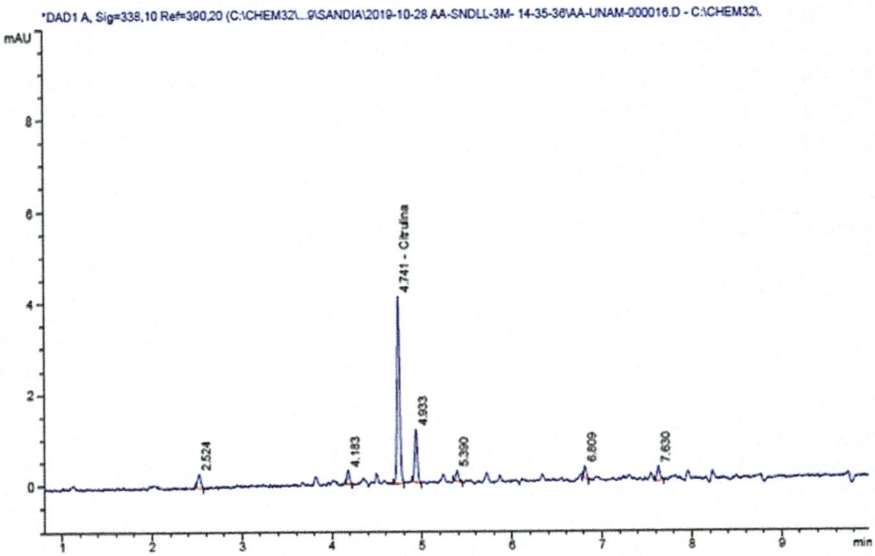

Fig. 2 Retention time of citrulline present in the mesocarp of maduro watermelon

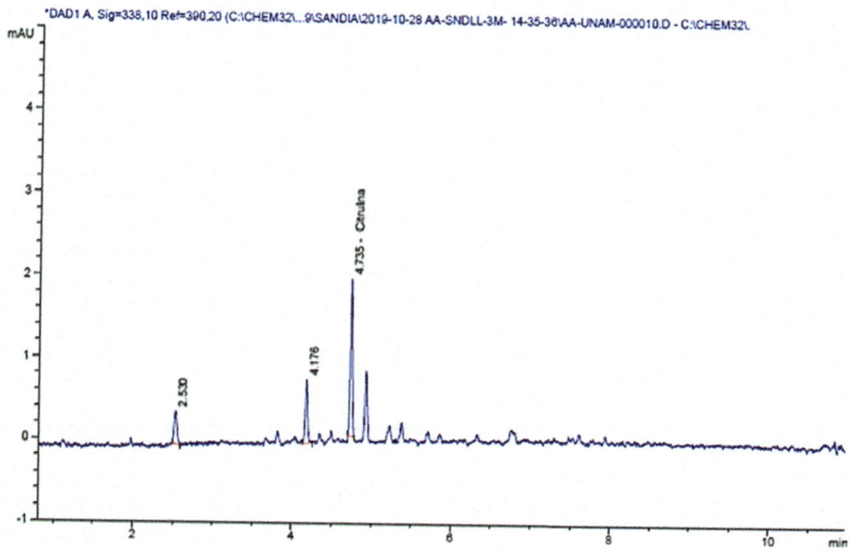

Fig. 3 Retention time of citrulline present in the mesocarp of watermelon in the pinto state

The retention time is defined as the time elapsed between the injection of the sample and the appearance of the maximum response, and the graph observed the "mAU" called m absorbance units, with respect to the elapsed time expressed in minutes. The highest citrulline content present in the sample of maduro watermelon mesocarp was recorded at minute 4.741 (highest peak in the graph). Given this retention time, a measurable signal is obtained on the y-axis (m absorbance units axis) by the diode array detectors of the equipment, and this signal is automatically processed by the HPLC software and expresses the value equivalent to the amount of amino acid contained in the sample. The time of retention of citrulline present in the mesocarp of watermelon in the pinto state is found in Fig. 3.

The time of retention of citrulline present in the mesocarp of watermelon in Verde state is found in Fig. 4.

Comparison of absorbances between treatments. In the following image, the three treatments are observed, which present a retention time of 4.7 min; nevertheless, they present different absorbances, being the mesocarp in mature state the one that registers greater absorbance (8.34122mAU*s) value much higher than of the pinto mesocarp (3.70746 mAU*s) and of the Verde mesocarp (2.96872 mAU*s) as found in Fig. 5.

Calculations of citrulline content (mg/100 g.) of treatments are found in Table 4.

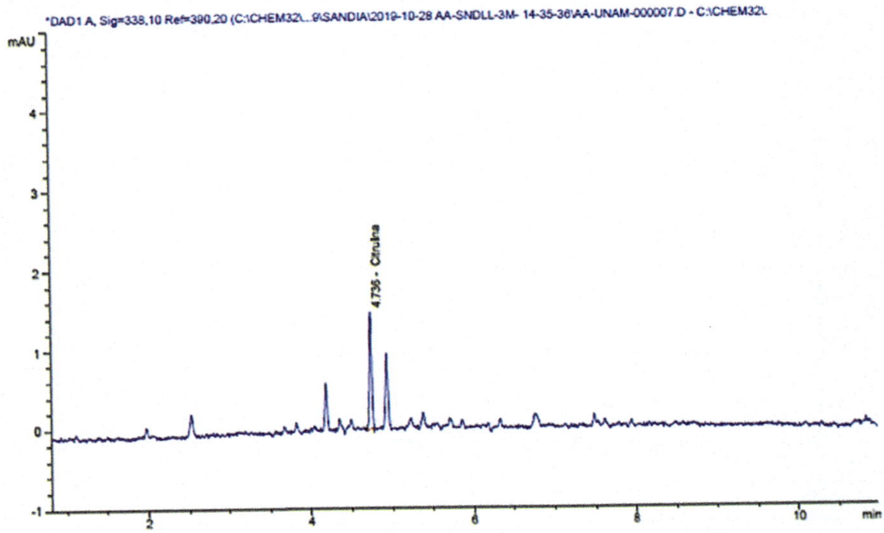

Fig. 4 Retention time of citrulline present in the mesocarp of watermelon in the Verde state

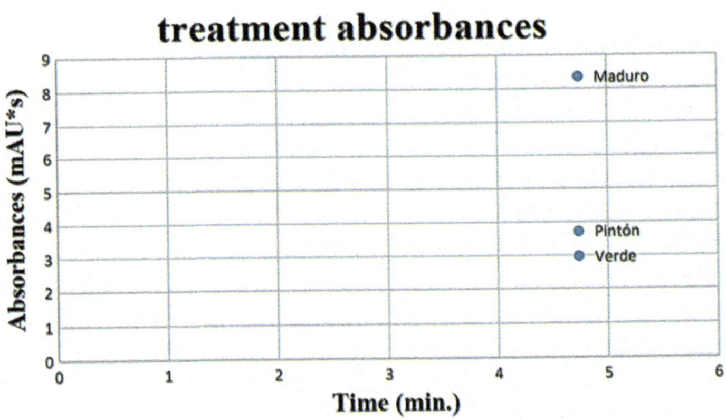

Fig. 5 Absorbance of treatments as a function of time

5.3 Statistical Treatment of Results

The analysis of variance shows the probability value "Pv." of the test "Fc" which is less than 0.01; with this, result confirms the mentioned hypothesis of the present research, and it is assumed that the ripening stage of watermelon influences the citrulline content in a highly significant way (Pv <0.01) as found in Table 5.

It was possible to determine which means are significantly different from others, it is observed that the greatest difference of means is presented between treatment T1 and T3, this means that, with respect to the Verde mesocarp and maduro mesocarp of

Table 4 Conversion of units in the analysis results

Treatment	Citrulline ug/ml	Capacity in 5 mL HCl 1 N	Citrulline ug/5 ml	Citrulline mg/5 ml	Weight of the sample (g)	Citrulline mg/100 g
Verde	291.4057	5	485.6762	0.4857	1.2670	38.333
Verde	291.6915	5	486.1525	0.4862	1.2670	38.370
Verde	291.7387	5	486.2312	0.4862	1.2670	38.377
Pintón	368.6435	5	614.4058	0.6144	1.3075	46.991
Pintón	368.0248	5	613.3746	0.6134	1.3075	46.912
Pintón	368.733	5	614.5549	0.6146	1.3075	47.002
Maduro	851.1638	5	1418.6064	1.4186	1.3438	105.567
Maduro	851.5217	5	1419.2028	1.4192	1.3438	105.611
Maduro	851.6807	5	1419.4678	1.4195	1.3438	105.631

Table 5 Analysis of variance of treatments

Source	S. C	g.l	Square middle	Fc	Pv
Model 1	8033.75	2	4016.87	2,981,843.34	0.0000
Maturity	8033.75	2	4016.87	2,981,843.34	0.0000
Waste	0.00808267	6	0.00134711		
Total	8033.76	8			
R2	99.9999%				

watermelon, and there is a highly significant difference in terms of citrulline content they present; there is also significant difference between the means of treatments T2 and T3, we refer to the watermelon mesocarp and pinto, and finally, there is significant difference in smaller proportion between the means of treatments T1 and T2, that is, between the Verde and pinto mesocarp of watermelon. The observed means exhibit a statistically significant difference at 95% confidence level as found in Table 6.

Confirmed in the Tukey test at 95%, it observed the significant difference that exists when comparing the values of citrulline among the Verde, pinto, and maduro mesocarp of watermelon as found in Fig. 6

Table 6 Tukey's method for multiple comparisons between treatments

Maturity	Cases	Media	Homogeneous groups
Verde	3	38.36	a
Pintón	3	46.9e683	b
Maduro	3	105.603	c

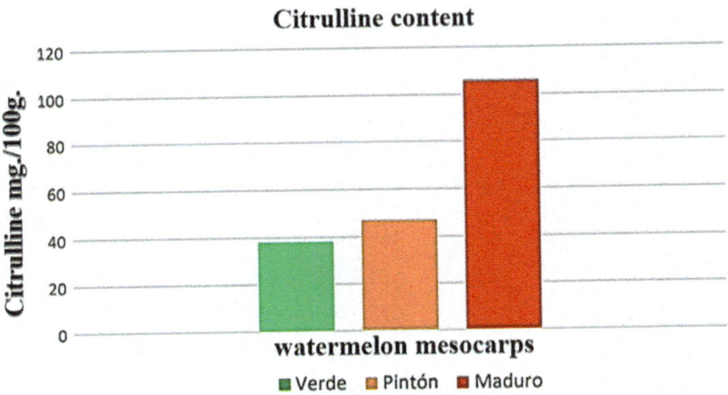

Fig. 6 Citrulline content of watermelon mesocarp under study

6 Discussions and Conclusions

As expressed by Rimando and Perkins in 2005, "The highest value in citrulline content, which shows a citrulline content in the mesocarp of maduro watermelon with values of 1.2–3.1 mg/g" [3]. From this study, it is evident that the mesocarp of maduro watermelon contains significant values of citrulline, whose beneficial potential in the human body can be exploited by giving it a proper use and prevent it from being wasted by simply discarding the watermelon rind.

The citrulline content in the mesocarp of watermelon is highly reflected according to the maturity stage. Watermelon with maduro mesocarp presents the best results in terms of citrulline levels (105.603 mg/100 g) compared to the other treatments under study. According to the comparison of means of Tukey, it is concluded that, from the pinto stage to the mature stage, the levels of citrulline in the mesocarp are high in contrast to the Verde stage. Based on the results obtained, it is concluded that the mesocarp of the watermelon produced in Nueva Requena offers a high content of citrulline, which should be used to combat the deficit of arginine in the human organism, contributing to the benefits offered by this amino acid and avoiding waste and contamination by organic solid waste.

References

1. Duran R et al (2014) Obtaining L-citrulline extract from watermelon rind using pressurized liquid. s.l.: Revista Facultad Nacional de Agronomía, pp 694–696
2. National Institute of Statistics and Informatics (2019)*Compendio Estadístico*. Ucayali: s.n, pp 257–258.
3. Rimando A, Perkins P (2005) Determination of citrulline in watermelon rind. s.l. J Chromatography 196–200
4. Collins J et al (2007) Watermelon consumption increases plasma arginine concentrations in adults 261–266.

5. Giaconi M (1989) Cultivo de hortalizas. Universitaria, Santiago, p 308
6. Peñaloza P (2001) Semillas de hortalizas: Manual de producción Valparaíso. s.l. Ediciones Universitarias de Valparaíso 161
7. Cruz P, Riofrío A (2012) *Obtaining amino acids from watermelon rind.* s.l.: University of Guayaquil 58
8. Crenn P, Cynober L (2010) Effect of intestinal resections on arginine metabolism: practical implications for nutrition support. s.l. Current opinion in clinical nutrition and metabolic care 65–69
9. Bahri S et al (2013) Citrulline: From metabolism to therapeutic use. s.l.: Nutrition 479–484
10. Wu G Amino acids: Metabolism, functions, and nutrition. s.l. Amino Acids I:1–17
11. Henderson J (2010) Improved Amino Acid Methods using Agilent ZORBAX Eclipse Plus C18 Columns for a Variety of Agilent LC Instrumentation and Separation Goals. s.l.: Agilent Technologies 16
12. Google Maps (2020) Geographical location of the National University of San Antonio Abad del Cusco (online). Cusco: s.n

Computational Pharmaceutical Chemistry and Analytical Chemistry

Molecular Docking Studies of *Coronavirinae* Spike Proteins with Different Vertebrate Receptors (ACE2, APN, DPP4)

A. C. Aquino and L. L. Tayo

1 Introduction

The subfamily *Coronavirinae* include coronaviruses of evolutionary, pathogenic, and biological importance as they have great genetic diversity, cause a broad spectrum of immune responses, and exhibit a complex receptor recognition process [1]. Coronaviruses generally contain 30 kb of positive-strand RNA, making them the largest of the currently identified RNA viruses [1]. They are enveloped by spike proteins and are pleomorphic but usually spherical [2]. The spike protein is an important component of the virion as it is responsible for binding with important molecules that either initiate viral reproduction, inactivate the virus completely, and trigger a series of immune responses [1, 2]. Currently, the highly mutated coronavirus spike protein may serve as a unique identifier for the classification of various coronaviruses [3]. Further discussion on the relationship of its amino acid sequence and its molecular behavior is required as reported by previous studies [6]. Hence, the spike protein and its role in coronavirus pathogenesis continue to constitute an active area of research [1, 6].

The path of coronavirus emergence is largely unknown [1]. Previous studies have shown crucial evidence that interspecies transmission of coronaviruses through intermediate hosts and/or subsequent adaptability to new hosts is key to coronavirus diversity [1, 4, 7, 8]. Additionally, ineffectiveness of some coronaviruses to infect humans in spite of rampant CoV-infected animals is also observed [5]. Most importantly, immediate emergence of SARS-CoV-2 variants is detected which are

---

A. C. Aquino
School of Graduate Studies, Mapua University, Intramuros, Manila, Philippines

L. L. Tayo (✉)
School of Chemical, Biological and Materials Engineering and Sciences, Mapua University, Intramuros, Manila, Philippines
e-mail: lltayo@mapua.edu.ph

© The Author(s), under exclusive license to Springer Nature Singapore Pte Ltd. 2023
S.-M. Chen (ed.), *Proceedings of 10th International Conference on Chemical Science and Engineering*, Springer Proceedings in Materials 21,
https://doi.org/10.1007/978-981-19-4290-7_12

more antigenic and exhibit greater transmissibility compared from the original virus [10, 11]. The ability of coronavirus spike proteins to utilize diverse receptors with different orthologues is one factor being considered. In humans, it is important to identify potential intermediate hosts to determine which animals are susceptible to possibly transmit and provide necessary mechanisms toward amplification of these coronaviruses [1, 2].

In this study, computational approaches were used to elucidate coronavirus spike protein interactions that influence coronavirus evolution and its complex virology. Specifically, the classification of coronaviruses was obtained from phylogenetic analysis of its spike proteins using maximum likelihood. Then, the most suitable interactions of coronavirus spike proteins belonging to different genus were determined from molecular docking of whole spike proteins with the ACE2, APN, and DPP4 receptors found in vertebrates using shape complementarity approach. Additionally, docking of SARS-CoV-2 variants may provide further insights on their higher transmissibility and antigenicity. In summary, this study showed the molecular basis of how coronaviruses use different cell receptors as modes of viral entry and ultimately, the susceptibility of various animals to accommodate coronaviruses that have not been experimentally proven yet to exist in these species.

2 Research Methodology

2.1 Data Gathering

The protein database of the National Center for Biotechnology Information (NCBI) was the source of homologous coronavirus spike protein sequences using Protein–Protein BLAST tool. The cell receptors used in this study are ACE2, APN, and DPP4 from human (*Homo sapiens*), bat (*Rhinolophus ferrumequinum*), cat (*Felis catus*), chicken (*Gallus gallus*), and rat (*Rattus norvegicus*).

2.2 Molecular Docking

Coronavirus classification was obtained after multiple sequence alignments of its spike proteins using multiple sequence comparison by log-expectation (MUSCLE) and maximum likelihood as statistical inference tool in MEGA X. The robustness of the tree was supported by 1000 bootstrap replicates [12]. The tree was rooted using an outgroup composed of toroviruses and bafinivirus spike proteins. Known and unknown tertiary structures of coronavirus spike proteins and receptors are obtained through the Protein Data Bank and predicted through homology modeling by PHYRE2 web server, respectively. Then, structures are cleaned and prepared using the DockPrep tool in Chimera.

Briefly, protein–protein interactions are understood as a solute in a solvent (e.g., water) involving heat transfer; hence, the laws of thermodynamics may be applied here [13]. The sum of all the interactions involved in molecular conformation change is the free binding energy E in kcal/mol. The more negative E value is predicted for a particular protein interaction, the more chance that the interaction is predicted to occur in nature as well as the spontaneity of the interactions [13]. A SPF-based docking software called Hex is chosen as the molecular docking strategy particularly the shape complementarity approach. Several of its advantages include using rotational correlations which reduce computation times up to two orders of magnitude compared to 3D Cartesian FFT utilized in other programs and the production of high quality predictions of docking compared to similar software involved in the Critical Assessment of PRedicted Interactions (CAPRI) [14].

2.3 Results Analysis

The binding energies in kcal/mol were obtained for each coronavirus spike protein-receptor complex.

3 Results

In this study, the amino acid sequences of the spike proteins of 50 members of the *Coronavirinae* subfamily with the latest variants of SARS-CoV-2 (Alpha, Beta, Gamma, Delta, Epsilon) were used for phylogenetic analysis with 6 *Torovirinae* as outgroup [15]. Specifically, *Torovirinae* is the other branch of the *Coronavirinae* family having spike proteins similar to coronaviruses [15]. Results showed that the addition of an outgroup defined the order of coronaviruses wherein from the root of the tree, *Alphacoronavirus* is the first genus to emerge followed by *Deltacoronavirus*, *Gammacoronavirus*, *Betacoronavirus*—Group C, *Betacoronavirus*—Group B, and lastly the most mutated, *Betacoronavirus*—Group A. The pairwise percentage identity and similarity of organisms in each group showed that all of the other genera excluding the genus *Betacoronavirus*—Group B have percent identity and similarity values of less than 50% with SARS-CoV-2. Most surprisingly, the genus *Gammacoronavirus* has the second to the lowest percent identity and similarity with SARS-CoV-2 with the outgroup composed of *Torovirinae* exhibiting the worst homology with SARS-CoV-2.

Molecular docking of coronavirus spike proteins with three widely expressed cell receptors namely ACE2, APN, and DPP4 from different animals containing these receptors was performed. Preferred animal species and cell receptors forming the most suitable interactions with notable coronaviruses in each classification are shown in Table 1. Furthermore, these results are compared to known animal hosts and preferred cell receptors based on current literature. Results showed that additional cell

Table 1 Notable coronaviruses per genus and their preferred animal species and cell receptors as calculated by Hex

Genus	Notable coronaviruses	Top animal species	Literature	Ref.	Top cell receptor	Literature	Refs.
BetaCoV-A	Bovine CoV	Bat	Cow	[14]	DPP4	O-acetylated sialic acids	[14]
	Human CoV-OC43	Human, cat	Human	[30]	ACE2	O-acetylated sialic acids	[30]
	Rodent CoV	Bat	Rodent	[14]	DPP4	CEACAM1a	[14]
	Porcine hemagglutinating encephalomyelitis CoV	Bat	Pig	[20]	DPP4	O-acetylated sialic acids	[30]
	Human CoV-HKU1	Bat	Human	[20]	DPP4	O-acetylated sialic acids	[30]
BetaCoV-B	Severe acute respiratory CoV	Cat	Human	[14]	ACE2	ACE2	[14]
	Severe acute respiratory CoV-2, Wuhan	Cat	Human	[14]	ACE2	ACE2	[14]
	Severe acute respiratory CoV-2 Alpha	Chicken	Human	[14]	ACE2	ACE2	[31]
	Severe acute respiratory CoV-2 Beta	Chicken	Human	[14]	ACE2	ACE2	[31]
	Severe acute respiratory CoV-2 Gamma	Chicken	Human	[14]	ACE2	ACE2	[31]
	Severe acute respiratory CoV-2 Delta	Chicken	Human	[14]	ACE2	ACE2	[31]
	Severe acute respiratory CoV-2 Epsilon	Cat	Human	[14]	ACE2	ACE2	[31]
BetaCoV-C	Middle East respiratory syndrome CoV	Bat, cat	Human	[14]	DPP4	DPP4	[14]
	Tyclonycteris bat CoV HKU4	Cat	Human	[14]	ACE2	DPP4	[14]
GammaCoV	Beluga whale CoV	Human	–	–	ACE2	–	–
	Bottlenose dolphin CoV	Bat	–	–	DPP4	–	–

(continued)

Table 1 (continued)

Genus	Notable coronaviruses	Top animal species	Literature	Ref.	Top cell receptor	Literature	Refs.
DeltaCoV	Porcine DeltaCoV	Cat	Pig	[14]	ACE2	APN	[14]
AlphaCoV	Feline CoV	Cat	Cat	[21]	APN	APN	[30]
	Canine CoV	Cat	Dog	[21]	DPP4	APN	[30]
	Human CoV-229E	Human	Human	[21]	ACE2	APN	[30]
	Human CoV-NL63	Bat	Human	[21]	DPP4	ACE2	[30]
	Porcine respiratory CoV	Chicken	Pig	[21]	APN	APN	[14]
	Transmissible gastroenteritis CoV	Cat	Pig	[21]	APN	APN	[14]

receptors were recognized by members of *Betacoronavirus*—Group A, Tyclonycteris bat CoV HKU4 of *Betacoronavirus*—Group C, *Deltacoronavirus,* and *Alphacoronavirus.* Interestingly, members of *Betacoronavirus*—Group B have similar affinity toward different orthologues of ACE2 receptors only (Table 1). In terms of host species, *Betacoronavirus* produced the most variant data except for human CoV-OC43, whereas SARS-CoV-2 variants recognized additional receptors with high affinity such as chicken receptors for Alpha, Beta, Gamma, and Delta variants and feline receptors for the Epsilon variant. MERS-CoV displayed high affinity toward bat and cat receptors, and Tyclonycteris bat CoV HKU4 displayed high affinity toward cat receptors (Table 1). Results for feline CoV and human CoV-229E showed affinity toward similar host receptors when compared to available literature, whereas additional animal hosts were recognized by canine CoV, HCoV-NL63, porcine respiratory CoV, and transmissible gastroenteritis CoV (Table 1). Lastly, cat was also observed as the preferred animal host of porcine DeltaCoV which is different from published data (Table 1).

4 Discussion

The hunt for natural reservoirs and potential intermediate hosts of coronaviruses is growing rapidly, and this important relationship is guided primarily by viral taxonomy. Most importantly, its interpretation is almost always influenced by bats being the widely known natural reservoir of coronaviruses [16]. However, other species are certainly underrepresented as potential hosts of coronaviruses. In this study, results for *Betacoronavirus* are aligned with previous literature [16]; hence, the chosen method provides comparable results among different coronaviruses and different variants of each coronavirus. Addition of newer coronaviruses is also provided in this study [17]. Most interestingly, majority of the members of *Betacoronavirus* are herbivores which indicate that the dietary preference and habit might be factors to consider in elucidating evolutionary relationships of coronaviruses other than the diversity and mobility of natural host species [15].

The prevalence of coronaviruses is constantly a challenge due to their molecular virology which contributes to the broad spectrum of illnesses associated with the infection [1, 2, 15]. Currently, biological assays determine the viral behavior of a limited number of coronaviruses [2, 7, 8]. However, such assays are time-consuming and require expensive reagents which may be avoided by a computational approach. Future recommendation for this study may include taking into account the location of cell receptor expression in the body. Briefly, the ACE2, APN, and DPP4 are ectopeptidases that are present in many animals due to their catalytic activities and are among the receptors heavily involved during coronavirus infection [18]. Findings may explain how coronaviruses are observed to jump from species to species and exhibit zoonotic transmission as previously discussed [2]. Lastly, the current study also hopes to contribute in establishing a kind of framework based on the different structural relationships that may be elucidated from coronaviruses toward potential viral surveillance [16].

Basis for molecular interaction of SARS-CoV-2 variants with different vertebrate receptors was also discussed for the first time which provides insights on these highly transmissible and antigenic variants [9, 11]. Such variants require urgent attention due to their immediate emergence and limited knowledge for their observed behavior [9]. Human ACE2 as the primary receptor of SARS-CoV-1 and SARS-CoV-2 has been discussed [18]; however, distinct characteristics in terms of cell receptor and host species were also observed. This pattern of affinity is only observed in SARS-CoV-2 variants, and this provides support on currently available data that SARS-CoV-2 recognizes multiple orthologues of ACE2 receptor [18] and further strengthens the possibility that multiple receptor recognition is the significant factor for the high transmissibility and antigenicity of these variants [9, 19].

5 Conclusion

In this study, classification of coronaviruses was inferred through phylogenetic analysis of coronavirus spike proteins using maximum likelihood. Molecular docking analysis of coronavirus spike proteins as well as SARS-CoV-2 variants (Alpha, Beta, Gamma, Delta, Epsilon) with ACE2, DPP4, and APN receptors of different animals (human, bat, cat, rat, chicken) using a SPF-based protein docking software was performed. Molecular basis of the different spike-receptor interactions for all coronaviruses as well as variants of the novel coronavirus was also provided. Findings of this study may provide intuitive information on coronavirus prevalence, cross-transmission, and immediate emergence of SARS-CoV-2 variants.

Acknowledgements This work was supported by the Engineering Research and Development for Technology scholarship by the Department of Science and Technology of the Philippines.

References

1. Colina SE, Serena MS, Echeverría MG, Metz GE (2021) Clinical and molecular aspects of veterinary coronaviruses. Virus Res 297
2. Hartenian E, Nandakumar D, Lari A, Ly M, Tucker JM, Glaunsinger BA (2020) The molecular virology of coronaviruses. J Biol Chem 295(37):12910–12934
3. De Albuquerque LP, Siqueira Patriota LL de, Gonzatto V, Pontual EV, Guedes Paiva PM, Napoleão TH (2020) Coronavirus spike (s) protein: a brief review on structure-function relationship, host receptors, and role in cell infection. Adv Res 21:116–124
4. Sohrab SS, Azhar EI (2020) Genetic diversity of MERS-CoV spike protein gene in Saudi Arabia. J Infect Public Health 13(5):709–717
5. Zhou Z et al (2021) Phenotypic and genetic characterization of mers coronaviruses from Africa to understand their zoonotic potential. Proc Natl Acad Sci USA 118(25)
6. Li F (2016) Structure, function, and evolution of coronavirus spike proteins. Annu Rev Virol 237–61
7. Terada Y, Matsui N, Noguchi K, Kuwata R, Shimoda H, Soma T, Mochizuki M, Maeda K (2014) Emergence of Pathogenic Coronaviruses in Cats by Homologous Recombination between Feline and Canine Coronaviruses. PLoS One 9(9)
8. Liang Q, Zhang H, Li B, Ding Q, Wang Y, Gao W, Guo D, Wei Z, Hu H (2019) Susceptibility of chickens to porcine deltacoronavirus infection. Viruses 11(6):1–13
9. Harvey WT et al (2021) SARS-CoV-2 variants, spike mutations and immune escape. Nat Rev Microbiol 19(7):409–424
10. Tegally H et al (2020) Emergence and rapid spread of a new severe acute respiratory syndrome related coronavirus 2 (SARS-CoV-2) lineage with multiple spike mutations in South Africa. medRxiv 2
11. Candido S et al. Genomics and Epidemiology of a Novel SARS-CoV-2 Lineage in Manaus, Brazil. 2021;1–45.
12. Kumar S, Stecher G, Li M, Knyaz C, Tamura K (2018) MEGA X: molecular evolutionary genetics analysis across computing platforms. Mol Biol Evol 35(6):1547–1549
13. Du X, Li Y, Xia YL, Ai SM, Liang J, Sang P, Ji X, Liu S (2016) Insights into protein-ligand interactions: mechanisms, models, and methods. Int J Mol Sci 17(2):1–34
14. Macindoe G, Mavridis L, Venkatraman V, Devignes MD, Ritchie DW (2010) HexServer: sAn FFT based protein docking server powered by graphics processors. Nucleic Acids Res 38(SUPPL. 2):445–449
15. Singh S (2013) [Viral Infections and Global Change]
16. Drexler JF, Corman VM, Drosten C (2014) Ecology, evolution and classification of bat coronaviruses in the aftermath of SARS. Antiviral Res 101(1):45–56
17. MacLachlan NJ, Dubovi EJ (2017) Chapter 24—coronaviridae. Fenner's Vet Viriology 435–461
18. Millet JK, Jaimes JA, Whittaker GR (2020) Molecular diversity of coronavirus host cell entry receptors. FEMS Microbiol Rev 14(October):1–16
19. Zhao X et al (2020) Broad and differential animal ACE2 receptor usage by SARS-CoV-2. bioRxiv 1
20. Graham RL, Baric RS (2010) Recombination, reservoirs, and the Modular Spike: Mechanisms of coronavirus cross-species transmission. J Virol 84:3134–3146. https://doi.org/10.1128/jvi.01394-09
21. Chen Y, Liu Q, Guo D (2020) Emerging coronaviruses: Genome structure, replication, and pathogenesis, J Med Virol 92:418–423. https://doi.org/10.1002/jmv.25681

Manufacturing Technology and Applied Mechanics

Fatigue Properties of Laser-Welded Laser Powder Bed Fusion Manufactured 316L Parts

Timo Rautio, Jarmo Mäkikangas, Jani Kumpula, and Antti Järvenpää

1 Introduction

Additive manufacturing (AM) has already gained footing in many industries including automotive, aerospace, dental and biomedical [1–3]. Laser powder bed fusion (LPBF) is the predominant method on the current market owing to the great mechanical properties and good surface quality compared to other printing methods [4]. In many cases, the manufacturing costs are still high compared to traditional methods, but many special features of the technique outweigh this disadvantage. Costs are very competitive when the need is for complex parts, and the technique also enables designing features that would not be possible in others [5]. Higher strength of the AM parts can also be utilized to reduce weight which is beneficial for the energy efficiency of the products.

Currently, LPBF has still some limitations, and usually, the orientation is limited in which a part can be printed due to the limitation on which angle the part can grow relative to the build direction without support [5]. This property can also limit the usage of some design features if their use requires a specific printing orientation. One possible tool to go around this problem is to manufacture the product from several parts and join them after the printing. This can be achieved with laser welding (LW).

T. Rautio (✉) · J. Mäkikangas · J. Kumpula · A. Järvenpää
Kerttu Saalasti Institute, University of Oulu, Pajatie 5, 85500 Nivala, Finland
e-mail: timo.rautio@oulu.fi

J. Mäkikangas
e-mail: jarmo.makikangas@oulu.fi

J. Kumpula
e-mail: jani.kumpula@oulu.fi

A. Järvenpää
e-mail: antti.jarvenpaa@oulu.fi

© The Author(s), under exclusive license to Springer Nature Singapore Pte Ltd. 2023

S.-M. Chen (ed.), *Proceedings of 10th International Conference on Chemical Science and Engineering*, Springer Proceedings in Materials 21,
https://doi.org/10.1007/978-981-19-4290-7_13

This work continues the authors' work on the research of laser-welded AM parts manufactured by LPBF. In terms of mechanical properties, LW is a suitable tool for joining AM parts with many materials including 316L [6], Ti6Al4V [7] and Inconel 718 [8]. Laser-welded 316L sheet material joints in terms of fatigue strength have been studied previously. PuchiCabrera et al. [9] studied the fatigue properties of gas metal arc welding process welded 316L under a pulsed arc mode and short circuit and also investigated the effect of shielding gas. Xiong et al. [10] investigated the very high cycle fatigue performance of the laser butt welded sheet metal 316L and found that the fatigue cracks initiated from internal pores of the weld, but still reached a fatigue strength comparable to the base material due to the convenient size of the pores. Laser shock processing can be used to increase the fatigue life of laser-welded sheet metal 316L as shown by Roman et al. [11]. However, there are no reports on open literature about the fatigue properties of laser-welded AM 316L, and this work addresses this. The microstructure of the material and the laser welds are investigated using optical microscopy, and tensile tests are used to verify the mechanical properties. In addition, the effect of LW on the fatigue life of the material is studied in flexural bending.

1.1 Material and Methods

Manufacturing of the Samples for Laser Welding Experiments An SLM 280 HL printer based on the LPBF technique was utilized for all samples manufactured for this work. Spherically shaped powder was supplied by Carpenter Additive of Carpenter Technology Corporation (UK) and had a particle size distribution in the range of 1545 μm and mean size of 31.7 μm. The composition of the powder batch used in this work is detailed in Table 1. For the bending fatigue experiments, rectangular-shaped specimens were printed in the size of 30 × 90 × 2 mm. Simultaneously, tensile was printed in the size of 120 × 20 × 2 mm. To reduce printing time and cost, the tensile was printed on 0° orientation, but the bending fatigue samples were 90°. All samples were printed on a single 280 × 280 mm platform using the same set of parameters which were as follows: laser power (P) of 200 W, speed (v) of 800 mm/s, hatch spacing (h) of 120 μm, layer thickness (t) of 30 μm and laser spot diameter 0.1 mm with a Gaussian laser profile. Applying these values to Eq. 1 results to an energy density of 69.4 J/mm^3.

$$E = \frac{P}{v \cdot h \cdot t} \tag{1}$$

Table 1 Chemical composition of the 316L powder

Fe	C	Si	Mn	P	S	Cr	Ni	Mo	N
Balance	0.02	0.54	1.24	< 0.005	0.004	16.72	12.14	2.38	0

Heat treatment (HT) for the stress relief purposes was carried out in a muffle furnace (Sarlin 1000HS436) under argon atmosphere after the printing. This process consisted of the following steps: heating at a rate of 10 °C/min to 600 °C and annealing for two hours followed by slow cooling to room temperature inside the furnace.

Laser Welding Bending fatigue properties of the laser-welded parts were determined for two series of LBPF manufactured 316L, the as built and HT conditions and compared to similar series without LW. A diode pumped continuous power ytterbium doped: yttrium aluminum garnet (Yb:YAG) laser with a maximum power of 4 kW, wavelength of 1030 nm and optical fiber diameter of 200 μm was utilized for the LW experiments. All welding was carried out using argon, which was directed right after the weld using a special shielding gas nozzle and at a flow rate of 30 l/min. Six-axis robot was utilized for the task, and the laser beam was directed perpendicular to the work piece. The following LW parameters were chosen for the welds: power of 2.2 kW, speed of 70 mm/s and focus at 1 mm below the work piece. After the LW, the bending fatigue samples and the tensile were machined to the appropriate shapes.

Characterization Instron universal material testing machine (Instron 8802) was used to conduct the tensile tests according to the standard SFSEN ISO 68921:2016 with a constant loading rate of 0.5 mm/min. For each case, the tests were repeated four times. Standard hourglass-shaped specimens with outer dimensions of 30 × 90 × 2 mm were manufactured.

On fatigue experiments, a reversed flexural bending fatigue machine by Carl Schenck was used with a stress ratio of $R = 1$ and frequency of 10 Hz. Stress amplitudes were used in the range from 200 to 700 MPa with runout of 2×10^6 w cycles. Both as built and HT conditions were calibrated separately with strain gages before testing. Fatigue tests were conducted in room temperature using air cooling when necessary.

Separate samples from all four conditions were prepared for metallographical analysis and analyzed with an optical microscope (Keyence VHX 2000 E) after polishing and etching in a 60% HNO_3 liquid using a voltage of 1.5 V for 10 s.

2 Results and Discussion

Microstructure of the Printed 316L

Microstructure of the printed material in as-built condition is presented in Fig. 1a. The fish-scale pattern formed by the melt pools introduced by the laser scanning tracks during printing is clearly visible. The build direction from right to left can also be identified from the figure. The microstructure mainly consists of columnar grains and cellular grains in very fine size. The cross section did not reveal significant porosity, but area of the image was chosen from an area where one lack of fusion fault is visible as black area. The high magnification in Fig. 1b reveals that the smallest grains are

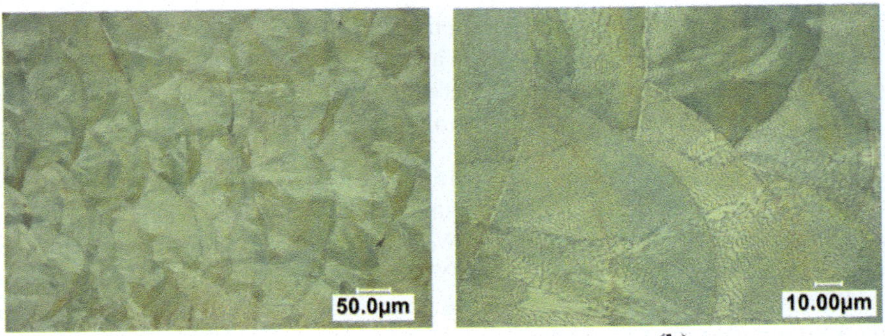

(a) (b)

Fig. 1 **a** Microstructure of the laser powder bed fusion manufactured 316L and **b** high magnification of (**a**)

only one micrometer in diameter. Some of the columnar dendrites can be seen to grow across the melt pool boundaries by epitaxial growth in the general direction of the heat flow during the printing. Microstructure after the HT for residual stress relief does not present any visible changes in the microstructure by optical microscopy analysis, so they are not presented.

Microstructure of the Laser-Welded 316L Both as built and HT conditions of LPBF manufactured 316L were laser welded, and the overall weld geometry is presented in Fig. 2. The weld is very narrow in general having a width of around 700 μm for the most part, but the top of the weld is wider at 1.2 mm. The weld center line is straight with no solidification cracking, and the weld is highly symmetrical with full penetration and good fusion. Fusion zone of the weld is presented in Fig. 2b showing the microstructural characteristics of the weld. The weld consists of mainly columnar dendrites and cellular grains. The size of both is largest near the weld center and reduced when approaching the base material. At the heat-affected zone Fig. 2c, no clear changes could be observed in the microstructure.

The Effect of LW and HT on the Mechanical Properties of 316L To measure the mechanical properties of the LBPF manufactured 316L, tensile testing was carried out for both as built and HT conditions. Both conditions were also laser welded, and the tensile tests repeated for comparison. Figure 3 presents these tensile test results in stress–strain curves for all four series, and the extracted values for yield strength (YS), ultimate tensile strength (UTS) and elongation at break are collected in Table 2. The results showed that the HT for stress relief reduced the YS by 6% from 495 MPa of the as-built condition to 465 MPa most likely due to dislocation rearrangement during the HT. On the other hand, a gain of 3.9% was recorded for UTS in exchange resulting to 653.3 MPa for the HT condition. The elongation at break was nearly identical for the two around 42%. Applying LW reduced the YS by 1.8 and 4.0% for as built and HT conditions, respectively. LW of the as-built condition did not result in any measureable difference in UTS, but in HT condition, 3.2% drop was recorded.

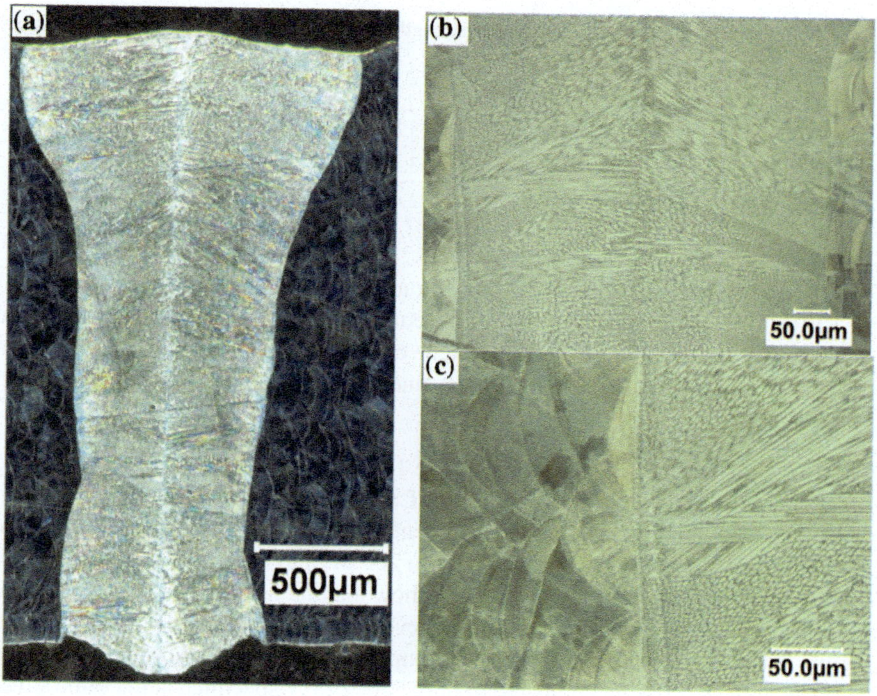

Fig. 2 **a** Full-weld profile of the laser-welded 316L, **b** fusion zone and **c** heat-affected zone

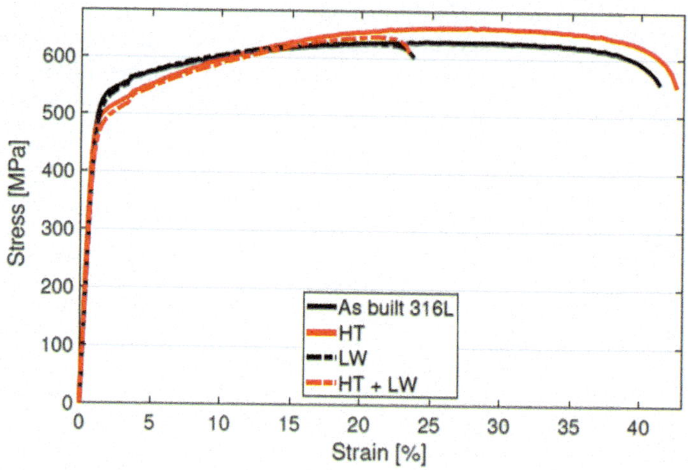

Fig. 3 Tensile test results for the LPBF manufactured 316L in as built and heat-treated conditions together with laser-welded versions of both

Table 2 Yield strength (YS), ultimate tensile strength (UTS) and elongation at break for the laser-welded 316L

	As-built 316L	HT	LW	HT + LW
YS (MPa)	495.3 ± 0.7	465.3 ± 2.9	486.2 ± 8.3	446.5 ± 9.3
UTS (MPa)	628.7 ± 1.0	653.3 ± 1.2	627.7 ± 2	632.2 ± 9.0
Elongation at break (%)	41.6 ± 0.2	42.4 ± 0.2	24.2 ± 0.5	24.3 ± 1.1

The results show that LW can be used without a significant effect on the strength of the LPBF 316L parts. However, in extreme conditions, the ductility is not at the same level as before the welding which should be considered. The weld has coarser grain structure compared to the base material reducing the strength and fracture strain in agreement with [12].

The Effect of LW and HT on the Fatigue Life of 316L Fatigue properties of the LPBF manufactured 316L in as built and HT conditions were studied with the effect of LW on both conditions using flexural bending machine. Based on the results on fatigue life on a stress amplitude range from 200 MPa to 700 MPa, the acquired SN curves are presented in Fig. 4. The results show that the fatigue limit of the material is not affected by the HT, but stays at the level of 200 MPa. However, at low cycle regime, the as-built condition is superior compared to the HT condition withstanding 44,000 cycles at 700 MPa compared to only 5700 of HT specimen at the same stress amplitude.

LW of the HT condition leads to further reduction in fatigue life at low cycle regime as at the 700 MPa level, the specimen endured only little over 1000 cycles. At lower stress levels, the fatigue strength becomes more and more similar with the non-welded material, and surprisingly, the fatigue limit was measured 25% higher at 250 MPa.

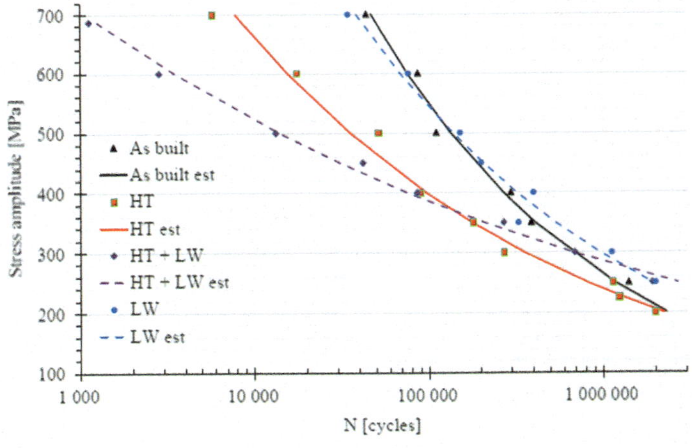

Fig. 4 Stress amplitudes and the corresponding fatigue lives for the as built, heat treated, laser welded and laser-welded after heat treatment conditions

Similar fatigue limit was reached with the laser-welded as-built material, but in this case, the results with higher stress amplitudes were nearly identical independent of the LW. These results suggest that the selected set of LW parameters works best for the as-built condition. It would be interesting to see if the results would improve if the HT was carried out only after the LW, but this was out of the scope of this study. Nevertheless, the results show that LW can be utilized without degrading the fatigue strength of the 316L material.

3 Conclusions

Laser welding (LW) of additively manufactured 316L was studied in this work focusing on the microstructure, mechanical properties and the fatigue life of the joints compared to the base material in as built and heat-treated (HT) conditions. The following conclusions can be drawn from this study:

- LW structure is characterized by columnar grains, and the weld is nearly free of any porosity. Symmetrical welds with very narrow heat-affected zones were achieved in all cases.
- Joint strength comparable to the base material in terms of yields strength and ultimate tensile strength was achieved in as-built condition of the material. In addition, minor reduction was observed in HT condition.
- The fatigue life of the base material is not affected by the LW joint in as-built condition, but nearly identical results were achieved by flexural bending fatigue testing. On the other hand, low cycle strength of the HT condition was reduced with the studied LW parameters.

Acknowledgements The authors would like to acknowledge the financial support received from the Council of Oulu Region and the European Union (European Regional Development Fund) for the "Hybridi" project.

References

1. Leal R, Barreiros FM, Alves L, Romeiro F, Vasco JC, Santos M, Marto C (2017) Additive manufacturing tooling for the automotive industry. Int J Adv Manuf Technol 92(58):1671–1676
2. Attar H, EhtemamHaghighi S, Soro N, Kent D, Dargusch MS (2020) Additive manufacturing of low-cost porous titanium-based composites for biomedical applications: advantages, challenges and opinion for future development. J Alloys Compd 827:154263
3. Froes F (2019) Additive manufacturing for the aerospace industry. Elsevier, Amsterdam, Netherlands
4. Kok Y, Tan XP, Wang P, Nai MLS, Loh NH, Liu E, Tor SB (2018) Anisotropy and heterogeneity of microstructure and mechanical properties in metal additive manufacturing: a critical review. Mater Des 139:565–586

5. Gibson I, Rosen D, Stucker B (2015) Additive manufacturing technologies. Springer, New York
6. Rautio T, Mäkikangas J, Jaskari M, Keskitalo M, Järvenpää A (2020) Microstructure and mechanical properties of laser welded 316L SLM parts. Key Eng Mater 841:306–311
7. Rautio T, Hamada A, Mäkikangas J, Jaskari M, Järvenpää A (2020) Laser welding of selective laser melted Ti6Al4V: microstructure and mechanical properties. Mater Today: Proc 28:907–911
8. Rautio T, Mäkikangas J, Kumpula J, Järvenpää A, Hamada A (2021) Laser welding of laser powder bed fusion manufactured inconel 718: microstructure and mechanical properties. Key Eng Mater 883:234–241
9. PuchiCabrera ES, SayaGamboa RA, La BarberaSosa JG, Staia MH, IgnotoCardinale V, BerríosOrtiz JA, Mesmacque G (2009) Fatigue life of AISI 316L stainless steel welded joints, obtained by GMAW. Weld Int 23(10):778–788
10. Xiong ZH, Ma XF, Qi XY (2019) Very high cycle fatigue behaviour of the 316L weldment fabricated by laser buttwelding. IOP Conf Ser: Mater Sci Eng 538:012025
11. Roman IB, Tierean M, Ocaña JL (2013) Effects of laser shock processing on 316L stainless steel welds. J Optoelectron Adv Mater 15:121–124
12. Zhang R, Buchanan C, Matilainen V, DaskalakiMountanou D, Ben Britton T, Piili H, Salminen A, Gardner L (2021) Mechanical properties and microstructure of additively manufactured stainless steel with laser welded joints. Mater Des 208:109921

Comparative Analysis of FE Modeling Techniques for Single-Lap Multi-column Composite Bolted Joints

Valerio G. Belardi, Pierluigi Fanelli, and Francesco Vivio

1 Introduction

Aeronautic and aerospace composite assemblies present several bolted joints. Their widespread diffusion is mainly motivated by the reliability and demounting capability of this connection technique. Consequently, for the design of complex composite structures, the assessment of multi-bolt joints is a determinant activity to guarantee the overall structural integrity. The finite element method is usually utilized to determine the bolt-load distribution in order to verify the most loaded bolt-holes. Performing a FE simulation of multi-bolt composite joints can be demanding, from both the modeling and the computational perspectives. In fact, a comprehensive analysis should include bolts preload, bolt-hole radial clearances and nonlinear effects, namely friction and contact between plates.

In this regard, the numerical analysis methods for composite bolted joints employ the finite element method and can be classified in three-dimensional and simplified approaches. The realization of a high-fidelity solid model is discussed in Ref. [1] in order to define guidelines for the choice of elements order, mesh density and for the modeling of Hertzian contact and bolt-hole clearance. In addition, the strategy for an accurate stress analysis is reported in Ref. [2], including the effect of radial clearance. Solid modeling is also proficiently employed to model the damage progression produced by different kinds of failure modes [3, 4]. Therefore, refined results can

V. G. Belardi (✉) · F. Vivio
Department of Enterprise Engineering, University of Rome Tor Vergata, Via del Politecnico, 1, 00133 Rome, Italy
e-mail: valerio.belardi@uniroma2.it

P. Fanelli
Department of Economics, Engineering, Society and Business Organization, University of Tuscia, Largo Dell'Università, 01100 Viterbo, Italy

© The Author(s), under exclusive license to Springer Nature Singapore Pte Ltd. 2023
S.-M. Chen (ed.), *Proceedings of 10th International Conference on Chemical Science and Engineering*, Springer Proceedings in Materials 21,
https://doi.org/10.1007/978-981-19-4290-7_14

be obtained with 3D models. Anyway, the main limitation in their employment is related to the time-consumption.

Subsequently, fast and accurate analysis approaches help to speed up the preliminary design process. Reference [5] describes a simplified approach made up of shell elements to reproduce the plates, a beam element for the modeling for the bolt and rigid contact elements. This model was improved and extended to multi-bolt joints Liu et al. [6]. Reference [7] outlines a semi-analytical model for eccentrically loaded multi-material joints. Then, Verwaerde et al. [8] presented a connector element, with low computational demand, working in both the axial and tangential directions.

Here, a simplified FE modeling methodology of the bolted region is utilized to obtain a precise design method with low simulation time requirements, the composite bolted joint element (CBJE) [9–11]. This method employs a group of finite elements, and their stiffness characteristics are obtained from the analytical model of bolted joint. Specifically, the elastic properties of the plate portions around the bolt are transferred to a group of custom radial beam elements. In addition, a beam element replaces the bolt-shank, and two spring elements simulate the friction and clearance effects. These features allow to combine low computational requirements with a rigorous derivation of the bolted region elastic behavior.

The performances of this FE tool are successfully compared with a solid model for the simulation of multi-column single-lap composite bolted joints with nine bolts. The CBJE results demonstrate high accuracy and limited computational requirements.

2 FE Modeling Approach

The FE modeling strategy utilizes the composite bolted joint element (CBJE), a group of basic finite elements with tailored elastic properties; the structure of the CBJE is shown in Fig. 1a. Additionally, the CBJE replaces a part of a pre-existing FE shell model in the position of the bolted joint. The elements composing the CBJE present a small amount of degree of freedoms if compared with analogous three-dimensional FE models, and their elastic properties are theoretically derived. These characteristics allow the simulation of the bolted region with limited computational effort and high accuracy. Consequently, the CBJE is particularly suitable for the structural analysis of multi-bolt joints.

Specifically, the CBJE can be split into the following components:

Sub-elements. Two groups of stiffness matrix elements with beam shape replace the region of composite plates surrounding the bolt. These elements connect the central node i on the bolt axis to the peripheral nodes j of the peripheral elements bordering the CBJE, see Fig. 1b. The sub-elements are structurally equivalent to the composite bolted joint analytical model, i.e., an angular portion of the laminate comprising the bolt. The theoretical solution of the reference model exploits the Ritz method to solve the principle of virtual displacements, and the description is reported in Refs.

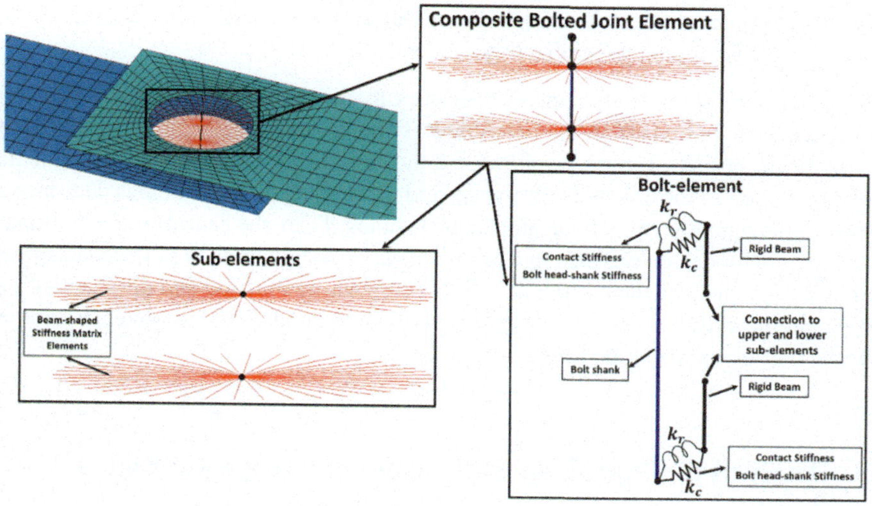

(a)

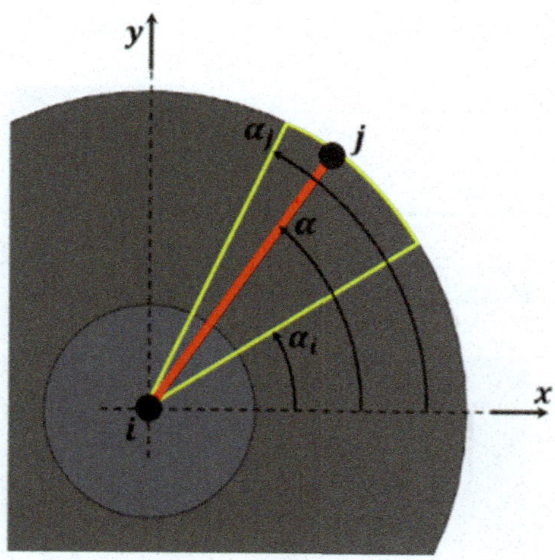

(b)

Fig. 1 **a** Configuration of the CBJE [11]. **b** Beam-shaped stiffness matrix element utilized to model a circular sector [11]

[12–15]. In particular, the properties of the beam elements are customized so as to be equivalent in stiffness with the *i*th angular region of the analytical model, limited by the circumferential coordinates α_i and α_j. Further details about the beam-shaped stiffness matrix elements are outlined in Refs. [11, 16]. Their evaluation is formulated as the ratio of the resultant force or moment that acts on the edge of the angular sector and the corresponding generalized displacement.

Bolt-element. A set of elements is necessary to simulate the elastic terms deriving from the bolt. In particular, the bolt-shank is simulated with a beam element. Then, a nonlinear spring element k_c allows for the modeling of the friction force of the plates, as well as the bolt-hole clearance, and the bearing displacement is determined by the contact of the bolt-shank and the plate holes. Then, the bearing deformation of the holes is modeled considering a beam on elastic foundation [17]. A second spring element k_r considers the rotational stiffness connecting the head and the shank of the bolt and the elastic support of the plates to the bolt head. The complete description of the bolt-element can be found in Ref. [11].

3 Analysis of Single-Lap Multi-column Composite Bolted Joint

The suitability of the CBJE methodology for the structural analysis of single-lap multi-column joints is tested considering a nine bolts joint subdivided in three columns. The geometrical characteristics of the joint are shown in Fig. 2. The joint is simply supported on the right edge of Plate 1 and clamped on the left edge of Plate 2; it undergoes a traction test with a load $F = 180$ kN. The thickness of the

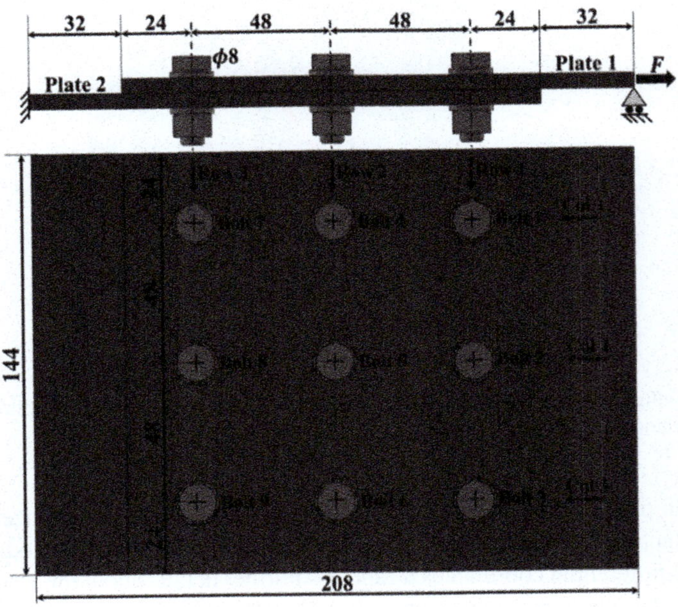

Fig. 2 Geometry of the single-lap multi-column composite bolted joint

composite plates is $t = 5.20$ mm, and they are obtained with quasi-isotropic stacking sequence: $[45/0/-45/90]_{5s}$. The unidirectional fiber reinforced layers feature the elastic properties reported in Ref. [11]; the layers thickness is $t_{\text{lay}} = 0.13$ mm. The nine bolts have a diameter $d_{\text{bolt}} = 8$ mm and are made up of titanium with Young's modulus $E_b = 110$ MPa and Poisson's ratio $\nu = 0.3$. The bolts are torque-tightened with a preload stress $\sigma_{\text{pre}} = 250$ MPa. Thus, being the friction coefficient of the material plates $\mu = 0.36$, the force threshold value determining the plates relative slipping is $F_s = 40.72$ kN. Besides, radial clearance of the bolt-hole is $c = 10$ μm for all the bolts.

The traction test of the single-lap multi-column composite bolted joint is modeled with a refined 3D model, used as reference for the results and with a shell model including the CBJE for the simulation of the bolted regions; the FE models are presented in Fig. 3a. As regards the 3D model, the elements are eight-node layered elements with three DOFs per node, and contact elements are present at the mutual interfaces of the components. The shell model with the CBJEs features four-node layered shell elements, each node characterized by six DOFs. The characteristics of the CBJE are defined with reference to the results reported in Ref. [11]; therefore, the CBJEs are composed of $N = 40$ custom beam elements to obtain a high accuracy level.

Furthermore, the FE modeling techniques are compared in terms of: (1) computational burden, (2) overall joint stiffness and (3) bolt-loads distribution. Table 1 lists the number of nodes and elements making up the 3D and the CBJE models;

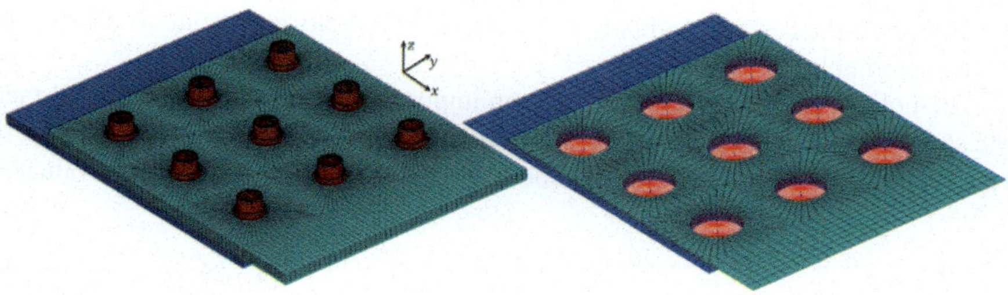

(a)

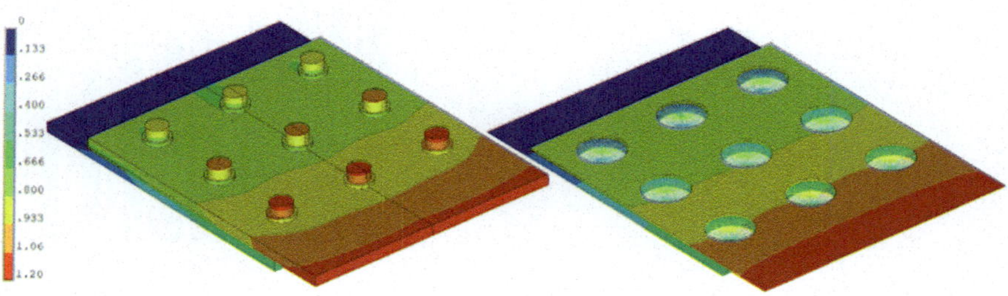

(b)

Fig. 3 **a** Single-lap multi-column composite joint: 3D model (left); model with CBJE (right). **b** Displacement u_x (mm) contour map along the traction direction

Table 1 Computational burden comparison of FE models

	3D FE model	CBJE FE model	Δ (%)
Nodes	341,312	8347	−97.55
Elements	360,384	8628	−97.61

the comparison demonstrates that the proposed simplified FE modeling approach drastically reduces the size of the model and, subsequently, the time required for the simulation. Also, despite the remarkable difference of computational resources necessary for the analysis, the CBJE preserves the reliability of the results. In fact, Fig. 3b reports the contour map of displacement u_x undergone by the specimens following to the application of the external load F: The displacement of the CBJE model is evaluated on the mid-surface, and it substantially coincides with the 3D model one.

As a further comparison, the stiffness prediction capabilities of the CBJE model are assessed through the load–displacement diagram of the joint, outlined in Fig. 4a, where the curves obtained with both modeling approaches are reported. It should be noted that the curves are in very close agreement. This aspect further proves the capabilities of the CBJE technique for the analysis of single-lap multi-column joints. Referring to the joint behavior, two stages of the load–displacement curve can be identified. The first one, with higher slope, is bounded by the origin of the axis and the point with force F_s. During this stage, the friction force does not allow reciprocal sliding of the plates. The second stage begins once the static friction force threshold value F_s is reached. Subsequently, the plates begin to slide reciprocally, the radial clearances are gradually regained, and the bolts contribute to bear a part of the external load.

The bolt-loads distribution represent an important result to be evaluated; in fact, the strength assessments are realized on the most loaded bolt. In this regard, Fig. 4b reports a histogram comparing the amount of load shared by each bolt and evaluated

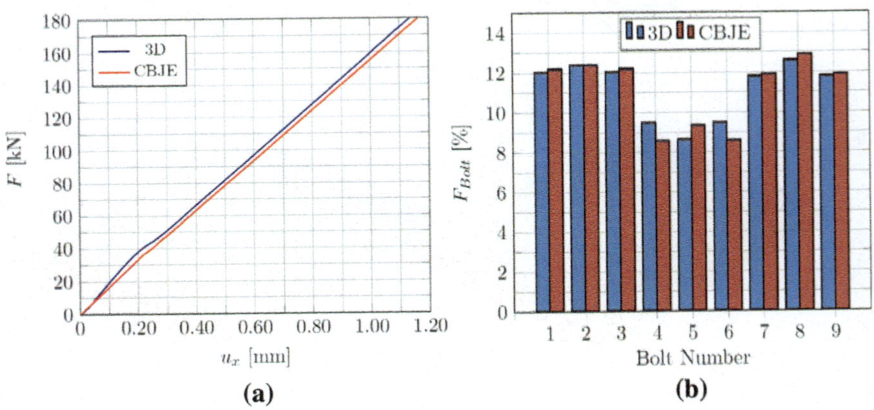

Fig. 4 **a** Load–displacement curve. **b** Histogram of bolt-loads ratio

with both the 3D and CBJE models. The differences of load estimation between the two modeling techniques are very limited. Additionally, it should be noted that the first and the third bolt rows carry higher portions of external load in respect to the central one, whereas Bolt 8 is the most loaded joint. Nevertheless, the external load is rather uniformly distributed among the bolts; in fact, the percentage differences of bolt-loads are below 4%.

4 Conclusions

This work evaluates the capabilities of a custom FE modeling strategy, the composite bolted joint element, for the analysis of single-lap multi-column composite bolted joints. The CBJE is a custom finite element with tailored stiffness properties, based on analytical derivation that is suitable for fast and reliable simulation of the bolted region.

Specifically, the traction test of a single-lap nine bolts, three-column joint is investigated. The results of a shell model comprising the CBJEs were compared with the ones of a reference 3D model. Firstly, it is found that the two modeling techniques strongly differ in terms of computational burden; i.e., the CBJE model allows for an important reduction of elements and nodes and thus of the simulation time. Additionally, the accuracy of the two models is comparable: The load–displacement curves are in very close agreement as well as the predicted bolt-loads distribution.

Finally, the main benefit obtainable with the usage of the CBJE is an important reduction of the computational time together with high accuracy level.

References

1. McCarthy MA, McCarthy CT, Lawlor VP, Stanley WF (2005) Three-dimensional finite element analysis of single-bolt, single-lap composite bolted joints: part I—model development and validation. Compos Struct 71:140–158
2. Egan B, McCarthy CT, McCarthy MA, Frizzell RM (2012) Stress analysis of single-bolt, single-lap, countersunk composite joints with variable bolt-hole clearance. Compos Struct 94:1038–1051
3. Mandal B, Chakrabarti A (2018) Numerical failure assessment of multi-bolt FRP composite joints with varying sizes and preloads of bolts. Compos Struct 187:169–178
4. Cheng X, Wang S, Zhang J, Huang W, Cheng Y, Zhang J (2017) Effect of damage on failure mode of multi-bolt composite joints using failure envelope method. Compos Struct 160:8–15
5. Gray PJ, McCarthy CT (2010) A global bolted joint model for finite element analysis of load distributions in multi-bolt composite joints. Compos B Eng 41:317–325
6. Liu F, Yao W, Zhao L, Wu H, Zhang X, Zhang J (2020) An improved 2D finite element model for bolt load distribution analysis of composite multi-bolt single-lap joints. Compos Struct 253:112770
7. El Masnaoui W, Daidié A, Lachaud F, Paleczny C (2021) Semi-analytical model development for preliminary study of 3D woven composite/metallic flange bolted assemblies. Compos Struct 255:112906

8. Verwaerde R, Guidault P, Boucard P (2021) A non-linear finite element connector model with friction and plasticity for the simulation of bolted. Finite Elem Anal Des 195:103586
9. Belardi VG, Fanelli P, Vivio F (2018) A novel composite bolted joint element: application to a single-bolted joint. Procedia Struct Integrity 12:281–295
10. Belardi VG, Fanelli P, Vivio F (2019) FE analysis of single-bolt composite bolted joint by means of a simplified modeling technique. Procedia Struct Integrity 24:888–897
11. Belardi VG, Fanelli P, Vivio F (2021) Theoretical definition of a new custom finite element for structural modeling of composite bolted joints. Compos Struct 258:113199
12. Belardi VG, Fanelli P, Vivio F (2018) Elastic analysis of rectilinear orthotropic composite circular plates subject to transversal and in-plane load conditions. Compos Struct 199:63–75
13. Belardi VG, Fanelli P, Vivio F (2018) First-order shear deformation analysis of rectilinear orthotropic composite circular plates undergoing transversal loads. Compos B Eng 174:107015
14. Belardi VG, Fanelli P, Vivio F (2021) On the radial bending of shear-deformable composite circular plates with rectilinear orthotropy. Eur J Mech A Solids 86:104157
15. Belardi VG, Fanelli P, Vivio F (2021) Ritz method analysis of rectilinear orthotropic composite circular plates undergoing in-plane bending and torsional. Mech Adv Mater Struct 28:963–979
16. Belardi VG, Fanelli P, Vivio F (2021) Analysis of multi-bolt composite joints with a user-defined finite element for the evaluation of load distribution and secondary bending, Compos B Eng 227:109378
17. Kou J, Xu F, Xie W, Zhang X, Feng W (2018) A theoretical 4-stage shear model for single-lap torqued bolted-joint with clearances. Compos Struct 186:1–16

Mechanical Properties of Laser-Welded Ultra-high-strength Stainless Steel Epoxy Foam-Filled Simple Panel Structure

Mikko Hietala, Markku Keskitalo, and Antti Järvenpää

1 Introduction

Transport-related emissions are among the most concerns in the context of climate change. Society's demands for lower traffic emissions are forcing the design of lighter vehicle structures. Laser-welded steel panels are one solution to reduce the weight of vehicle structures. Laser-welded steel panels have become more common due to their excellent mechanical properties and lightweight [1]. One major advantage of steel panels is that the use of panels can increase the stiffness of the structure. Other benefits can also be achieved by using panels. For example, studies have shown that the lowest natural frequency of the steel panels can be much higher than that of traditional welded structure made of sheet metal [2].

Laser-welded steel panels made of ultra-high-strength (UHS) steels in particular have very good mechanical properties. When the steel panel is manufactured of UHS steel, the thickness of the sheet metal can be reduced without compromising the strength of the panel.

Steel panels are usually consisting of face plates which are separated by a core. Laser-welded steel panels can be manufactured from many different cores, such as corrugated, honeycombs, pyramidal truss. Bending strength of laser-welded steel panels have been recently investigated (Kananen et al. [2] and Sun et al. [3]).

M. Hietala (✉) · M. Keskitalo · A. Järvenpää
Kerttu Saalasti Institute, University of Oulu, Pajatie 5, FI-85500 Nivala, Finland
e-mail: mikko.hietala@oulu.fi

M. Keskitalo
e-mail: markku.keskitalo@oulu.fi

A. Järvenpää
e-mail: antti.jarvenpaa@oulu.fi

© The Author(s), under exclusive license to Springer Nature Singapore Pte Ltd. 2023

S.-M. Chen (ed.), *Proceedings of 10th International Conference on Chemical Science and Engineering*, Springer Proceedings in Materials 21,
https://doi.org/10.1007/978-981-19-4290-7_15

The lightest steel panels are made of thin plates; therefore, traditional welding methods for panel manufacturing cannot be applied as they could cause heat distortions. Laser welding is also only welding method that makes steel panel manufacturing economically viable due to the high welding speed. The very low heat input of the laser welding is a big advantage due to the good dimensional accuracy of the panels can be achieved.

Studies have shown that filling steel panels with foam can significantly improve the mechanical properties of the panel [4–8]. Yan et al. [9] investigated three-point bending of sandwich beams with aluminum foam-filled corrugated cores. They found that foam-filling increased bending stiffness and peak load of the sandwich beams. Karttunen et al. [10] studied foam-filled steel sandwich beams under three-point bending for stiffness. Their investigation showed that the overall stiffness of the sandwich beam was three times that of the empty beam. Yan et al. [11] investigated compressive strength of sandwich panels with aluminum foam-filled corrugated cores. They found that foam-filling of the panel increased the compressive strength of the panel approximately 200%.

The simple panel structure (SPS) tested in this study was developed by an FMT research group at the University of Oulu. The SPS has been developed to provide the lightest possible structure, but at the same time, the strength of the structure is at a sufficient level. This study examines whether the mechanical properties of the SPS can be further improved by filling it with epoxy foam. The bending and compression strength of the laser-welded epoxy foam-filled SPS were evaluated.

2 Experimental Methods

2.1 Test Panel Manufacturing and Material

The SPS was made of 0.3 mm thick austenitic stainless steel AISI 301 (ASS). The ASS was supplied by Georg Nobitschek GmbH & Co. Mechanical properties of ASS can be seen from Table 1. ASS was cold-worked ultra-high-strength steel, and its yield strength (YS) was 1526 MPa. The chemical composition of the SPS panel material can be seen in Table 2. Blanks of the test material were cut with a guillotine for forming. The blanks were formed into C-shaped strips with press brake. Dimensions of the strip can be seen in Fig. 1a. The SPS is assembled so that the C-shaped strips are stacked inside each other, and each of the strips is the other way around as seen in Fig. 1b. C-shaped strips were joined by laser welding as seen in Fig. 1b. The SPS

Table 1 Mechanical properties for the SPS panel material

Material	YS	UTS	Elongation	Hardness
	(MPa)	(MPa)	(%)	(HV0.2)
ASS	1526	1644	5.0	511

Table 2 Chemical composition (%) of the SPS panel material

Material	C	Si	Mn	P	S	Cr	Ni	Mo	N
ASS	0.096	1.05	1.07	0.032	0.001	16.94	6.42	0.71	0.062

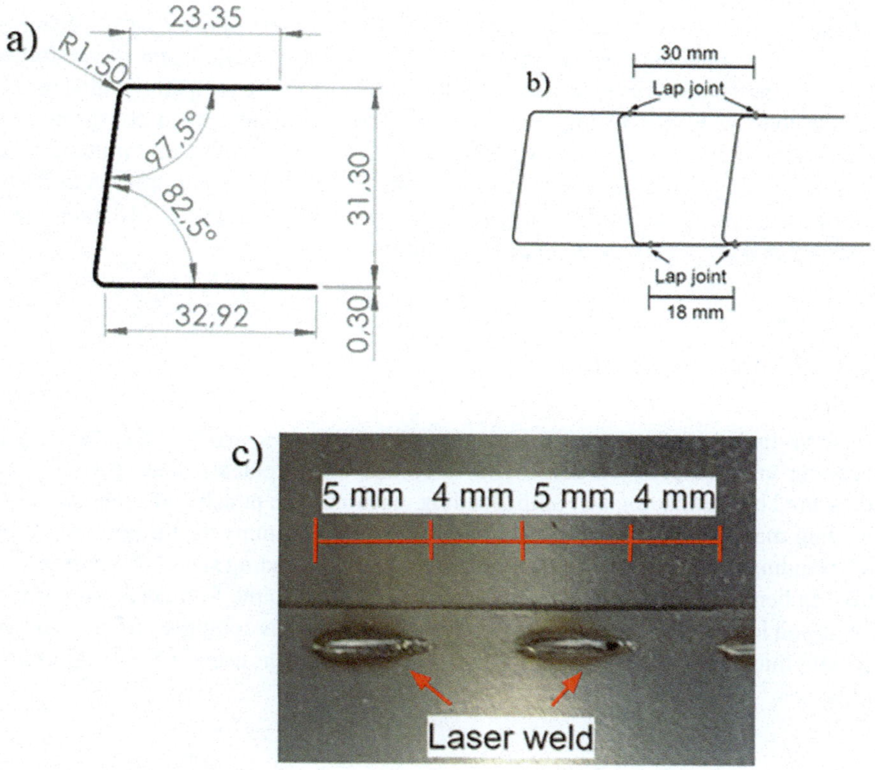

Fig. 1 **a** Dimensions of strip, **b** SPS assembly, and **c** Dimensions of the laser-welded intermittent lap joint

was filled with epoxy foam (Sicomin PB 170) with a nominal density of 170 kg/m^3. The epoxy foam was supplied by Kevra Oyj. The epoxy foam was cured at 40 °C for 12 h, so it had no effect on the mechanical properties of the ASS. The epoxy foam-filled SPS was coded as SPS-EF. The samples (Fig. 1) for the compression test were dimensions with length of 100 mm and width of 100 mm. The samples for the three-point bending tests were 300 mm long and 300 mm wide. The weight was 7.61 and 15.06 kg/m^2 for SPS and SPS-EF, respectively.

2.2 *Test Panel Laser Welding*

Yb:YAG disk-laser (Trumpf HLD-4002) with a maximum power of 4 kW was used in this paper for laser welding of the test panels. The focal distance of the laser optics was 300 mm, and the diameter of the laser beam was 0.3 mm. The focal point was adjusted 1 mm below the surface of panel. The beam quality of the laser was 8 mm × mrad. Argon shielding gas was used in the laser welding, and flow rate was 30 l/min. Intermittent weld was used for lap joints of the SPS, because for a thin-walled panel, it is a good weld due to low energy input. The intermittent weld consisted of a 5 mm long welds with a 4 mm spacing as seen in Fig. 1c. The intermittent weld was selected on the basis of previous experience due to low heat input in order to avoid deformations in the final structure [12]. Laser welding speed of 18 m/min and constant power of 4 kW were used for welding.

2.3 *Mechanical Testing*

A servo-hydraulic loading machine (Instron 8802) was used in the three-point bending and compression tests. The force and displacement of the panels were measured by the loading cell of the loading machine. The punch and supports of the loading machine were semi-circular steel bars ($R = 25$ mm) for the bending tests. The loading span between the supports was 250 mm. Loading speed of 1 m/min was used in bending tests. The load directions with respect to the SPS cores were transverse and longitudinal. In compression tests, the panel was compressed between two 20 mm thick steel plates. Loading speed of 0.5 m/min was used in the compression tests.

3 Results and Discussion

3.1 *Bending Tests*

Four different cases were studied by three-point bending experiments. Bending experiments were performed on SPS and SPS-EF. Both panel types were tested with the core's transverse and longitudinal in the bending direction. Results showed that the direction of the cores relative to the bending direction had significant effect on the bending strength of the SPS as seen from Fig. 2. When the cores were transverse in the bending direction, the bending strength of the SPS was 14.7 kN. With the cores longitudinal in bending direction, the bending strength of the SPS decreased significantly and was approximately 0.1 kN. Poor bending strength is due to the very thin face plate (0.3 mm) of the SPS. It was noticed that when the SPS cores were longitudinally orientated to the bending direction, SPS was prone to face yielding. SPS

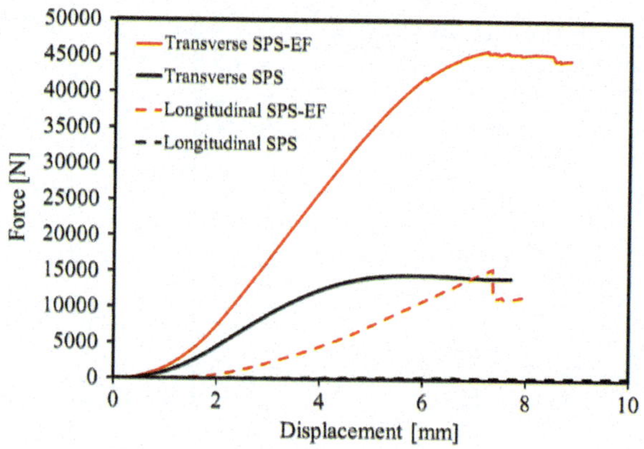

Fig. 2 Bending strengths of the SPS and SPS-EF

cores in transverse orientation to the bending direction, the SPS failure mode was also face yielding. There were large differences in the mechanical properties of the SPS depending on the load direction because SPS is not homogeneous. Järvenpää et al. [13] investigated bending strength of corrugated core steel sandwich panels. They found that sandwich panels had very significant anisotropy in mechanical properties depending on the direction of the cores with respect to the bending direction.

It was observed that filling the SPS with epoxy foam significantly improved the bending strength. As seen from Fig. 2, the bending strengths were 46.5 and 16.1 kN for SPS-EF with core orientations transverse and longitudinal to the bending direction, respectively. As seen from Fig. 2, load–displacement curve shows that SPS-EF cores longitudinal to the bending direction had sudden failure. This is due to SPS face plate buckling when the epoxy foam fractures.

3.2 *Compression Tests*

Compression tests were performed on two panel types SPS and SPS-EF. The purpose of the tests was to determine the benefits of epoxy foam under compressive loading. Comparison between SPS and SPS-EF showed that compressive strength of SPS-EF was significantly better than that of SPS as seen from Fig. 3. Peak compressive strength was 0.41 and 4.43 MPa for SPS and SPS-EF, respectively. Filling the SPS with epoxy foam increased the peak compressive strength by approximately 980%. Wang et al. [14] investigated foam-filling techniques to enhance mechanical behaviors of sandwich panels. They observed that polyurethane foam improved the compression strength of the panel from 0.596 to 5.14 MPa (~760%).

The compression strength of epoxy foam was 1.8 MPa. The linear summation of strengths of SPS and epoxy foam is 2.21 MPa which is considerably less than the strength of the SPS-EF (4.43 MPa). There is clear coupling effect based on foam-filling. Result is comparable to a study by Yan et al. [15]. They investigated effects

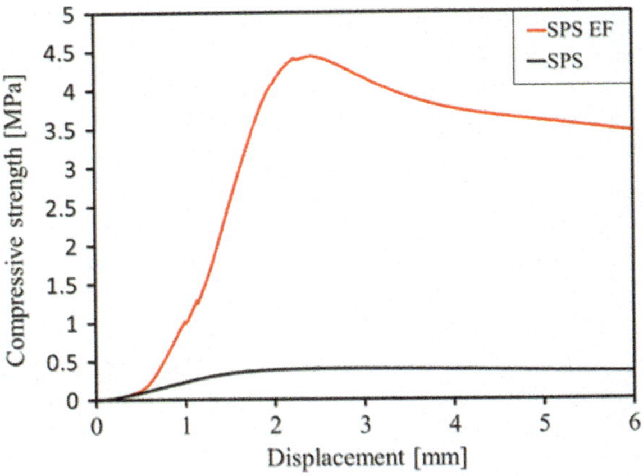

Fig. 3 Compressive strengths of the SPS and SPS-EF

of aluminum foam-filling on compressive strength of metallic sandwich panel. Their observation was that the strength of the foam-filled sandwich panel was higher than sum of the strengths of the foam and empty sandwich panel.

Figure 4 shows the deformation mode of SPS and SPS-EF under compressive load. The core of the SPS buckled under compressive loading as shown in Fig. 4a.

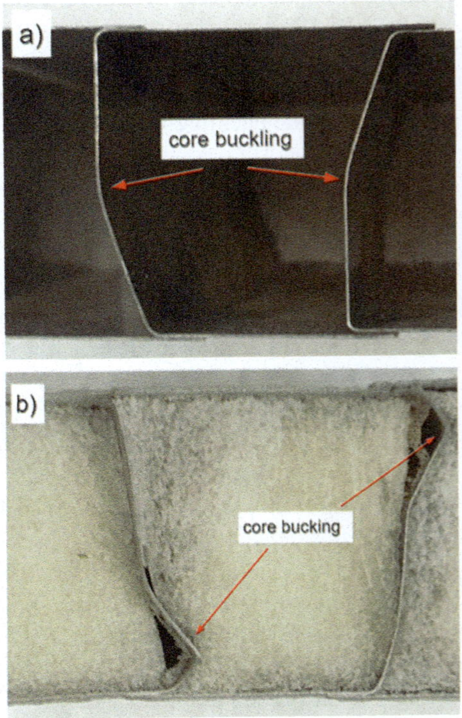

Fig. 4 Deformation mode of **a** SPS and **b** SPS-EF

In the SPS, the greatest deformation is directed to the center of the core, from which the core buckles. The deformation mode of the core changed with SPS-EF and the core buckled near the face plate as shown in Fig. 4b. The epoxy foam provides lateral support to the core, and buckling of the thin core requires significantly more compressive force. In the SPS-EF, the foam supports the core, causing the maximum torque to be generated at the ends of the core near the face plate and the deformation to occur at those points as seen in Fig. 4b. Yan et al. [15] observed that maximum core deflection of the empty sandwich panel was in each case larger than deflection of the foam-filled panel.

4 Conclusions

This paper investigated the bending and compression strengths of epoxy foam-filled simple panel structure (SPS-EF). Results were compared to unfilled SPS. The following conclusions may be drawn from the present investigation on mechanical properties of laser-welded UHS stainless steel epoxy foam-filled SPS:

- The bending strengths were 14.7 and 0.1 kN for SPS with core orientations transverse and longitudinal to the bending direction, respectively. SPS exhibits significant differences in the mechanical properties depending on the load direction due to SPS is not homogeneous. This should be taken into account in the design of the SPS structures.
- Load–displacement curves showed that SPS-EF had significantly higher bending strength than that of SPS. The bending strengths were 46.5 and 16.1 kN for SPS-EF with core orientations transverse and longitudinal to the bending direction, respectively.
- Filling the SPS with foam dramatically improved its compressive strength. Compressive strength of SPS-EF was approximately 980% better than that of SPS.
- There was a clear coupling effect based on foam-filling. The sum of the compression strengths of the SPS and epoxy foam was considerably less than the compression strength of the SPS-EF.
- Filling the SPS with foam changed the deformation mode of the core due to the fact that the foam supported the core laterally.

Acknowledgements The authors would like to acknowledge the financial support received from the European Union (European regional development fund), City of Nivala, Nivala industrial park Ltd, NIHAK Nivala-Haapajarvi region registered association. The industrial companies Wärtsilä Finland Oyj, SSAB Europe Oyj, HT Laser Oy and Randax Oy have participated in this research work.

References

1. Jiang XX, Li JM, Cao R, Zhu L, Chen JH, Wu YX, Li ZG (2014) Mater Sci Eng A 595:43
2. Kananen M, Mäntyjärvi K, Keskitalo M, Hietala M, Järvenpää A, Holappa K, Saine K, Teiskonen J (2015) Phys Procedia 78:202
3. Sun Y, Saafi M, Zhou W, Zhang C, Li H (2015) Mater Today: Proc 2:279
4. Cheng Y, Liu M, Zhang P, Xiao W, Zhang C, Liu J, Hou H (2018) Int J Mech Sci 145:378
5. Zu G, Lu R, Li X, Zhong Z, Ma X, Han M, Yao G (2013) T Nonferr Metal Soc 23:2491
6. Vălean C, Şoşdean C, Marşavina L, Linul E (2021) Mater Today: Proc 45:4166
7. M. Kazemi: Structures Vol. 29 (2021) p. 383
8. Taghipoor H, Eyvazian A, Musharavati F, Sebaey TA, Ghiaskar A (2020) Structures 28:424
9. Yan LL, Han B, Yu B, Chen CQ, Zhang QC, Lu TJ (2014) Mater Des 60:510
10. Karttunen AT, Kanerva M, Frank D, Romanoff J, Remes H, Jelovica J, Bossuyt S, Sarlin E (2017) Mater Des 115:64
11. Yan LL, Yu B, Han B, Chen CQ, Zhang QC, Lu TJ (2013) Compos Sci Technol 86:142
12. Hietala M, Järvenpää A, Keskitalo M, Jaskari M, Mäntyjärvi K (2019) Procedia Manuf 36:131
13. Järvenpää A, Lämsä J, Hietala M, Mäntyjärvi K (2014) Key Eng Mater 611–612:781
14. Wang H, Li S, Liu Y, Wang P, Jin F, Fan H (2021) J Build Eng 40:102383
15. Yan L, Su P, Han Y, Han B (2020) Metals 10:1670